THE RARE AND THE BEAUTIFUL

ALSO BY CRESSIDA CONNOLLY

The Happiest Days

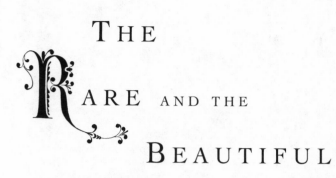

THE
RARE AND THE
BEAUTIFUL

*The Art, Loves, and Lives
of the Garman Sisters*

CRESSIDA CONNOLLY

ecco

AN IMPRINT OF HARPERCOLLINSPUBLISHERS

HarperCollins books may be purchased for educational, business, or sales
promotional use. For information, please write: Special Markets Department,
HarperCollins Publishers Inc., 10 East 53rd Street, New York, NY 10022.

FIRST EDITION

Designed by Claire Naylon Vaccaro

Library of Congress Cataloging-in-Publication Data

Connolly, Cressida, 1960–
    The rare and the beautiful : the art, loves, and lives of the Garman
sisters / Cressida Connolly.—1st ed.
        p.   cm.
    Includes bibliographical references and index.
    ISBN 0-06-621247-2
    1. Garman family.   2. Garman, Kathleen, 1901–1979.   3. Campbell,
Mary, 1898–1979.   4. Garman, Douglas, 1903–1969.   5. Wishart, Lorna,
1911–2000.   6. Epstein, Jacob, Sir, 1880–1959—Marriage.   7. Campbell,
Roy, 1901–1957—Marriage.   8. London (England)—Social life and
customs—20th century.   9. London (England)—Biography.   I. Title.

CT787.G37C66 2004
929'.2'0942—dc22
                                                                2004043211

04   05   06   07   08   BVG/RRD   10   9   8   7   6   5   4   3   2   1

To my sister and brothers

Sarah Bradbury

Simon Craven

Matthew Connolly

Contents

Preface

Early in the autumn of 2000 I went to see the New Art Gallery in Walsall, West Midlands, less than an hour from where I live. The gallery had been much in the news, partly because it was a spectacular new piece of modern architecture, and partly because it was home to a remarkable collection of art, the Garman Ryan Collection. Here were works by Monet, Van Gogh, Constable, Rembrandt, Degas, Rodin, Dürer, Cézanne, Burne-Jones. There were sculptures, votive objects, and vessels from Africa, Asia, South America, and the islands of the Pacific. Twentieth-century art was especially well represented, with works by Picasso, Modigliani, Matthew Smith, Augustus John, Lucian Freud, and Gaudier-Brzeska. At the nucleus of it all were a number of family portraits by the American-born sculptor Jacob Epstein.

From the catalogue I learned something of the personalities who had assembled the collection before giving it to the region. One was an heiress, the little-known American sculptor Sally

Ryan, and the other was Epstein's widow, the alluring and mysterious Kathleen Garman. Included in the catalogue was a vivid personal account by the Epsteins' daughter Kitty Garman.

This name rang a bell. I remembered that Kitty Garman and her first cousin Michael Wishart had been friends of my parents. Their world had overlapped with that of the Garmans and their children: Chelsea in the 1930s, nights at the Gargoyle Club in Soho and afternoons sitting for Augustus John, lunches with Aldous Huxley in the south of France before the Second World War. Later, when I spent time in the archive where Kathleen Epstein's books are housed, I found many books which I recognized from home. There were the baby blue covers of Scott Moncrieff's translation of Proust; an austere navy-and-gold set of Henry James; signed first editions of T. S. Eliot, Theodore Roethke, and Allen Tate; and the fading Chinese yellow binding of the Larousse *Mythology*, with its introduction by Robert Graves. Another link was that some of the Garmans had settled near the sea by the South Downs in Sussex, where I grew up.

I sent a letter to Kitty Garman to see whether I might obtain her permission to write something about her mother. At that point I had only an article in mind. Kitty Garman replied that she'd always thought that someone should write a book about her mother, and her mother's sisters. It turned out that Kathleen Garman had had six sisters, each of whom had led a remarkable life. The Garmans were strikingly beautiful, artistic, and flamboyant. Mary, the eldest, had married the poet Roy Campbell and become the lover of Vita Sackville-West before settling in the Mediterranean. Another sister, a willowy lesbian given to wearing an aviator's hat, had been possibly the only female lover of T. E. Lawrence (a story I was never able to verify, alas). One had endured terrible personal loss; one had worked for the Free French during the Second World War; and one had, according to her septuagenarian son, done nothing but fornicate. The youngest, Lorna, was the loveliest of all. She had married at sixteen, before becoming an inspiration to the poet Laurie Lee and the painter Lucian Freud. Then there were two brothers, Douglas and Mavin, both of whom had been members of the Communist Party; one a writer and the lover of Peggy Guggenheim, the other an extremely good-looking cowboy.

It seemed incredible that there hadn't yet been a book about this family. They had lived at the center of European literary and political life between the two world wars, numbering some of the greatest artists and writers among their husbands, friends, and lovers. They were very exotic, dark and tall and graceful, with huge, limpid eyes. They had been dazzling company, brilliant mimics who set out to enchant everyone they met, and generally succeeded. The Garmans had an intoxicating quality, and people felt their lives had been transformed by knowing them. Almost everybody who knew them called them "just magical."

But the Garmans disliked publicity and preferred to lead their lives out of the spotlight. In old age, Lorna even went so far as to burn all the love letters from her many admirers, so that no one else could read them. Douglas's widow pruned her husband's papers of any personal material before donating them to Nottingham University Library, and Kathleen's last companion also excised all but the most anodyne letters, notebooks, and diaries. Most of the Garmans would have been baffled and appalled by the current cult of celebrity and by the popular culture of which it forms a part. To them, *popular culture* would have been an oxy-moron. They lived the low life, sometimes, in Soho or Marseille, but they were always passionately high-minded: art and litera-ture and music were essentials. They were prepared to put up with extremes of poverty in the cause of Art, to sacrifice com-fort, the approval of society, friendships, even the happiness of their own children. Kathleen's letters show that she was a natural writer and she played the piano to concert standard, but she ceded her talent for Epstein's. Douglas's communism probably stopped him from fulfilling his potential as a poet and novelist. Mary and Lorna were both gifted artists, but they chose to de-vote their energies to love, and then to the love of God.

This book does not pretend to be a full biography of the Garmans. Instead, it tells stories of the four whose lives are most fully documented: Mary Campbell, Kathleen Epstein, Douglas Garman, and Lorna Wishart. I have tried to respect the privacy of their living children and have written about them only in order to shed light on their parents' characters.

In her old age, Kathleen Garman Epstein was approached by a young man who thought of writing her biography, an en-counter she described to a friend. "He asked if I had ever thought of writing my memoirs," she reported, "and I said 'never never never.' And then he said his worst utterance, 'Would you consider letting me write your life?' at which I just got up and said, 'It's time you went, now. I can't even discuss such a ridiculous proposition.' Words fail me!" Kathleen went on, with evident relish. "The mind boggles, in the famous words of the taxi driver. What muddy pitfalls one inadvertently steps into in search of the rare and the beautiful."

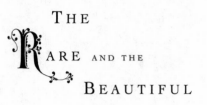

THE RARE AND THE BEAUTIFUL

Chapter One

THE BLACK COUNTRY

Walsall in the West Midlands has been called the ugliest town in the world. One visitor described it as looking like the worst of Ceausescu's Romania, only with fast-food outlets. It is notorious for its high incidence of mugging and its low property prices. Many of its shop windows are darkened by heavy steel bars. There is a pawnbroker and a mothy casino and a pedestrianized shopping street strewn with an urban confetti of cigarette butts and chewing gum.

The Garman Ryan Collection had lain all but forgotten in Walsall's old library, where it had been put on view to the public in 1974. But when it was removed to the specially built New Art Gallery in 1999, there was a spate of publicity as arts journalists and broadcasters saw the collection for the first time. Local renewal schemes followed on the heels of the gallery's success. The disused factories below the gallery began to be converted, the litter-strewn canal basin cleaned and restored. An impressive, architect-designed new bus station was built, and a shopping mall. Walsall was waking up.

The art collection at the heart of Walsall's improving image had been given to the town at the wish of one woman. Lady Epstein, née Kathleen Garman, was born only a handful of miles from Walsall. She and her eight brothers and sisters had been a most unusual family. They valued naturalness very highly; they barely disciplined their children; they spoke their minds. The sisters wore their hair straight and long when custom called for stiff permanent waves. They liked things to look effortless. Elaborate picnics appeared, as if out of nowhere, and their houses were models of elegant simplicity in which important and valuable drawings and paintings would be propped casually against the walls. They accepted the most extraordinary coincidences as nothing less than their due.

People fell in love with them. They were lovely to be in love with, passionate, generous, beautiful. They sent secret notes at midnight and left their pillows smelling of scent. They gave presents: books of poetry, music, wildflowers. They made dramatic entrances and exits, their arms full of lilies, haunting railway stations throughout Europe, intoxicating their lovers with sudden meetings and long good-byes. On his deathbed, a former lover finished his last letter to one of the sisters with "I do not forget you ever." To the poet Laurie Lee, Lorna Garman left an indelible mark on the rest of his life, an imprint of a "dark one, her panther tread, voice full of musky secrets, her limbs uncoiling on beds of moonlight."

They sought adventure, emotional altitude. Color mattered.

Their letters are full of it: the bright blue sky of the Italian Alps, the scarlet leaves of a persimmon tree, the light-saturated palette of Mediterranean France and Spain, the purple robes of a bishop at an abbey tea party, the rose-pink buildings of Tuscany, the magnificent vermilion of dahlias. To understand the Garmans, it is necessary to see that this world of color and intensity stood in sharp contrast to the dark, industrial region that they came from, in the shadow of the First World War.

Every night the sky was lit up by the flames of the blast furnaces down in the valley, and in summer the pale roses in the garden would be covered with tiny flecks of black. Soot fell like snow. Smoke from the smelting of iron stained the sky, while coal inked the earth beneath. Even the trees were darkened, and the rare black form of the peppered moth, *Biston betularia*, was believed to have become widespread because it could camouflage itself so well here. Not for nothing was this region, just to the northwest of Birmingham, called the Black Country. Until the late nineteenth century—and again during the First World War, because of the burgeoning need for munitions—the area was a hub of the iron and coal industries. There was so much iron locally that even the street curbs were made of it. Anchors and chains were sent all over the world, and the ironwork for the Crystal Palace, the Avon Suspension Bridge, Westminster Bridge, and Charing Cross railway station was made here. A local vicar wrote, apparently without irony, that the people round-about were "never more happy than when enveloped in a cloud of smoke, for then, though the rays of the natural sun be interrupted, the sun of prosperity gladdens its people."

It was here in the heart of the Black Country, to Oakeswell Hall in Wednesbury, that Walter Garman brought home his bride, in the late spring of 1897. Walter was tall and dark, with expressive eyes and arched brows. He wore a drooping mustache and a slightly melancholy look, like a Spanish don. He had already been disappointed in love when a local girl broke off her engagement to him, but his heart gradually opened to the sweetness of his new companion. Margaret Magill, Marjorie as she was called, was just twenty-one, while her husband was in his late thirties. The two had met while Marjorie was at school with three of Walter's four sisters. When she took up a post as a governess in Coventry, she went frequently to see her best friend, Mabel Garman, at Yew Tree House, Great Barr, the Garman family home, just east of Wednesbury. Mabel had three brothers, and the family rather expected Marjorie to lose her heart to one of them. But they were slightly taken aback that she chose Walter, the eldest son, as if she had blithely taken everyone's favorite chocolate when proffered the box. Much as they liked Marjorie, serene and pale and serious, with blue-gray eyes, fair skin, and long, chestnut hair, she was not a catch. Her mother was an impecunious widow, and the daughter, unlike the Garman sisters, was no beauty. But she was cultured and intelligent, gentle and kind and true. Several of her grandchildren would grow up to describe her as a saint. She read widely, loved Beethoven and the fugues of Bach, and she had a profound appreciation of the natural world. She was evidently a woman of outstanding sweetness of character.

During their courtship, Walter Garman read aloud to her from Shakespeare and Tennyson, as he would later read to their children during evenings in the nursery. The letters exchanged between the two in the months leading up to their wedding are ardent and tender: "I have just been [for] the sweetest walk there is round here, called Harvest Hill," wrote Marjorie early in the engagement, "which I connect so much with you, for it has been my favourite walk throughout that silent year, when I dared hardly confess to myself how much I loved you." Walter is her "beautiful lover" and a "beautiful knight," and she feels that they are "almost living in Heaven."

Walter's replies are no less affectionate. "The sense of incompleteness I have, being without you, my Beauty, teaches me so truly of the love I bear towards you. Throughout today, many a time, I have missed your smile, your voice, your many-sided presence..." He calls her his own darling, his precious girl.

Majar, my sweetheart, do you know I wondered so much whether after all events of the past, my sentiment would ever be kindled anew or whether my past experiences had blasted and killed my greatest feelings. Now, however, no such doubts remain, for your sweet and fresh simplicity has begotten in my inmost self a feeling of tenderness for *you*, different from and I fancy beyond what my earlier thoughts prompted—and which I long to maintain fresh and pure as you have caused it to be. I want my best life to be lived around you, encircling you in a lasting zone of love.

He found that the beauties of nature intensified his feelings: "I heard the bird sing, and thought so of you—I saw the new moon, just risen, with the Evening star, and wondered and gazed on them and thought of you, whether you saw them and sent forth your love to me..." On April 27, 1897, he sent her a touching postcard: "My own darling, I send this little note—the last I shall write to you as *Marjory Magill*—so I take this opportunity to send you a word while you bear that name. The gig is returning here and then I am driving up to see you on this the last day before your wedding—a momentous day, marking a new epoch. You are in my mind constantly—Oh, that I may help you realise all your sweetest hopes! Always my darling, yr own Walter Garman."

They married the next day at the parish church of Great Barr. The church still stands perched on the hill above a narrow lane with banks of cow parsley, an oasis of rustic charm in the endless urban sprawl of Greater Birmingham. It was a double ceremony, with Walter's sister Florence and her groom, Arthur Romney Green, who was later to become a renowned furniture maker in the Arts and Crafts style. "We went to the village church in a coach with postilions," Romney Green later recalled of the day. "A solitary shaft of sunlight fell at the altar."

Walter was, like his father and grandfather before him, a doctor. The family were of such local distinction that a street in Great Barr is named after them. Walter studied medicine at Heidelberg and Edinburgh, and in 1888 he had become medical officer for Wednesbury, for the annual salary of eighty-four pounds. He was personally responsible for the introduction of flush toilets to the area, an innovation that must have saved many thousands of people from typhoid and cholera. Walter was immensely popular, would help the poor without charge, and although he claimed never to use anything but aspirin and vinegar, was known as an extremely fine doctor. He delivered children, set broken bones, and would even operate on kitchen tables. He used to go out on his horse in all weather to assist at the accouchements of miners' wives, and he was always on the scene at pit accidents.

Walter was a churchwarden, a magistrate, and a keen member of the local Conservative Party. He rode to hounds. Such a man needed a house befitting his status, and in Oakeswell Hall he had it. The house was Jacobean, with rather gloomy Victorian additions, such as pebbledash and cosmetic half-timbering. There were a number of outbuildings, including one that Walter used as his surgery, all built of red brick, much grimed and blackened by smoke. There was a large walled garden shaded by tall trees, where he gave tennis parties, croquet parties, and teas on the lawn. There was a pony called Silver Bells for the pony cart, which the family had for everyday use, and a somber black brougham for special occasions.* Walter employed a lady's maid, a kitchen maid, a housemaid, a scullery maid, and a chauffeur, a butler, a cook, a gardener, and a housekeeper, Mrs. Fowl, who did the shopping every day. After the children were born, the staff expanded to include a nursery maid, Ada Newbould, and a governess, Miss Thomas.

Walter was a devoted husband. When he visited the Alps—*the* chic destination in the first decade of the twentieth cen-

---

* In 1916, to his great sadness, Walter's horses were requisitioned for the war. He soon acquired his first motorcar, but habits of a lifetime of horsemanship took a while to abate, and whenever he put his foot on the brake he would say, "Whoa."

tury—he sent long, affectionate letters home to Marjorie. "We have had a lovely time but it seems so horrid of me to be seeing all these beauties without my own M," he wrote. "Still darling you couldn't have come this time, could you?" This was because Marjorie had, in the summer of 1899, given birth to their second daughter, Sylvia. Their first, Mary, had been born in 1898. During the twelve years that followed, Marjorie was hardly ever not pregnant or nursing. By the time she had been married for ten years, she had produced six children. Walter, who always wrote poetry, commemorated their wedding anniversary with a verse and a ring:

And take to wear this emblematic ring
Six jewels add lustre to the central gem;
As thy six children, round thee clustering,
Enhance thy charms, which live anew in them.

There were to be nine children in all, three groups of three: Mary and Sylvia and Kathleen, who was born in 1901. Then came Douglas (a flag was flown from the highest post in Wednesbury in 1903 when he, their first son, was born), Rosalind in 1904, and Helen in 1906. Last came Mavin, in 1907; Ruth, in 1909; and Lorna, in 1911. Their cousins always knew them as the Naughty Nine.

Both Dr. Walter and Marjorie were passionately religious, and church played a large part in the children's upbringing. Church attendance was a constant (the doctor went to early mass every morning on horseback), and clerics often visited the

The Garmans in 1913. From left to right: Kathleen, Lorna, Mavin, Mary, Rosalind, Douglas, Ruth, Sylvia, Helen.

house. Douglas's godfather, Father Tudball, later went to minister in a leper colony. Canon Boddington of Lichfield Cathedral, a benign white-bearded man and close family friend, sat the children on his knee. Every morning, in the nursery before breakfast, the whole household would be led in prayer by the doctor. His air of moral authority was absolute, in sharp contrast to his gentle, somewhat remote, wife.

Mostly, the nine young Garmans had an idyllic childhood. There were trips to the mountains and the seaside in Wales, and picnics and golf and tennis. Douglas was taken fishing by his father on the banks of the Wye and the Welsh Dee. For years, until the family became too large, they would take a house, Lapley Hall near Cannock, where they spent their summers. There would be riding and croquet and beautiful aunts in flowery muslins, and in the evenings when the children were in bed, music. Lapley had once been a monastery, and the monks' fishponds remained, full of water lilies and with shallow steps leading down to the water. The children would float little arrangements of flowers and fallen blossoms and berries on the water, ready for their paternal grandmother, with her rustling skirts and parasol of ivory and lace, to come and judge which was the most beautiful and imaginative. Holidays were also taken at the country house at Abberley in Worcestershire to which Dr. Walter's parents, Dr. William and Mary Garman, had retired. The whole household would remove to the elder Garmans', complete with maids, horses, and endless wicker baskets and trunks.

Oakeswell mornings were given over to lessons in the schoolroom across the courtyard, before the children returned to the Hall for lunch. Walter himself gave two German lessons a week, and there was also a German master. Herr Füchner was regarded with a mixture of dislike and sympathy: he would rap the children sharply on their fingers if their attention strayed, but they took pity on him when they guessed that he was very hard-up, since he generally helped himself to all the milk and biscuits at midmorning break. Joe the gardener, a particular friend to Douglas, taught them how to ride their mother's old bicycle, incurring the wrath of Walter, since the tires were so flat that the rims of the wheels cut up the lawns. The girls were taught to play the piano by a Miss Thorrington, a tiny bird of a woman. Sometimes she would arrive at Oakeswell to be told that the girls had mislaid the key that locked the piano lid, so that there could be no lesson that day. But the piano could not often have been locked, for Kathleen was to become a highy talented musician. A Madame Bouvier came to conduct French conversation over nursery tea. Their father read aloud to them: prose on weekdays, poetry, as a treat, on Sundays.

In Wednesbury, the daily walk—from noon until one o'clock, come rain or shine—included visits to the painted barges on the canal, where the bargemen would sometimes invite the children in to warm themselves by the stove. The children especially enjoyed walking to a forge called Stonecross, where they would watch the horses being reshod. The bellows would roar and sparks leap as the white-hot metal was shaped and the glowing nails hammered. To the little Garman children, the smith seemed almost like a god, in his control of fire and metal and beast. At home in the afternoons, they played games of make-believe, dressing up as the knights of King Arthur and performing elaborate ceremonies presided over by Mary, who acted as high priestess. Or they pretended to be characters from *Uncle Tom's Cabin*, rescuing maltreated slaves from the snake-infested swamps of the vegetable garden. The firelight would flicker on the nursery wall at night, when the children were curled up in their beds.

Years later, Kathleen wrote about Oakeswell to a friend:

> I wish you could have seen it in its heyday, overrun with children and horses and pony and dogs and cats and rabbits and guinea pigs and cocks and hens—an old courtyard, surrounded

by the house itself and all around the stables, the schoolroom, the laundry, an old malt house we called the bogey hole where we had three unsafe storeys of dirty and dusty old buildings for hiding and playing in and the top storey of which we made in to our private chapel with frescoes painted by my sister Mary. When we were not in a religious mood we had feasts up there. In the Spring we spent whole days in a beautiful beechwood carpeted thickly with bluebells at Charlemont Hall which I believe is now a housing estate. And the whole of Oakeswell is a wilderness of weeds.[*]

There was a very different world outside the walls of Oakeswell, glimpses of which may have influenced the political convictions that some of the Garmans later adopted. They noticed that the altar boys at church were skinny and wore frayed and dirty clothes underneath their surplices. Returning from shopping trips in the brougham with their mother, they saw barefooted little boys darting among the horse-drawn traffic, selling newspapers. During a coal strike, Dr. Walter set up a soup kitchen for the miners' children in the large boiler room. At six o'clock each evening, a long queue of hungry children would tramp up the drive, each holding a bowl or mug, and range themselves around the banks of the tennis courts. The young Garmans helped the cook and housekeeper hand out hot broth and baskets of buttered rolls to the strikers' children.

The Garman boys were sent to prep school in Rugby, and at thirteen, both went on to Denstone, a minor public school in

---

[*] The house fell into disuse and was demolished in 1962.

south Derbyshire, while their sisters boarded at a girls' school in Abbot's Bromley in Staffordshire. Mavin hated school, but Douglas became head boy of his prep school and later distinguished himself in English and Rugby football. Mary, meanwhile, passed her higher school certificate with honors and fell in love with an attractive female painting teacher. The older girls cut a dash, even then. They wore swirling black cloaks instead of the usual dull tweed coats of the time, and black velvet ribbons in their hair. People thought they looked rather French. Some sixty years later, a cousin still remembered them as teenagers practicing their flirting, making pretend-eyes at him.

On her son's engagement, Walter's mother had written to Marjorie's about Walter: "I can promise you that she will have one of the most honourable, good-tempered, affectionate, high principled men that it is possible to meet with and I cannot see any reason that such a union can fail to be a happy one." The marriage does seem to have been very happy. But the assurance about Walter's good temper may not have been entirely true. He seems to have inherited a tendency to rage from his own father, and there are various accounts of him taking it out on his children. Walter was very insistent about standards of schoolwork and behavior. Douglas incurred a severe blasting for wasting money by taking a puncture on his bicycle to be mended, instead of repairing it himself. Ruth was to recall that she and Lorna, the two youngest, would hang on to their father's coattails to try to prevent Walter from beating the others. Lorna remembered occasions in the formal dining room when he served everyone else at the table but her. He would wait until they were all halfway through the meal, before saying, with feigned surprise, "Oh! You haven't got any food."

Although he was immensely popular for miles around, there were whispers in Wednesbury that the doctor suffered from mental problems. Many years later, Helen told Stephen Gardiner, the biographer of Jacob Epstein, that her father's rages had been the result of sexual frustration. "He was a cruel, bad-tempered, ferocious man who beat his children and locked them in cupboards," she said. "He beat them unmercifully—oh yes, the girls as well as the boys—for trifling offences, and chiefly, apparently, to rid himself of raging sexual desires that couldn't find fulfilment. Of course, they had to be strenuously repressed on account of his position. Otherwise he would have been struck off and ruined in no time." But Helen was a great one for exaggerating. Mavin remembered only one beating, when he and Helen were caught cheating at their math lessons.

Kathleen was, by common consensus, her father's favorite. It is easy, from contemporary photographs, to see why. She was dark and wistful, with wide cheekbones and slightly slanting eyes. She looked more like her aunt Mabel than any of the other children, and Mabel was Walter's favorite sister. Everyone remembered that Kathleen never got beaten, and some even seemed to think that Dr. Walter was so attached to her that he would sometimes stop her from going to school. She would later describe herself as "the little daughter who was his constant companion." He may have bullied Helen because she was awkward and perhaps cheeky, and Mary, because he thought she was too forward. But

Kathleen amused and charmed him. She was witty and clever, and she excelled at chess and the piano. As a party piece he would invite local friends to play chess, setting each man at a table with board and pieces before him. Then Kathleen, trained by her father and now under his proud gaze, would come into the room and go from table to table, winning every game. She could wrap her father around her little finger. Even so, she once admitted that she thought he "had madness in him."

Kathleen began to attend outside music lessons in 1917, probably at the nationally regarded Birmingham Midland Institute. Mary went on the tram with her sister to art classes in Birmingham. These two were already showing signs of independence and spirit. Kathleen used to send Helen and Mavin out to Wednesbury as couriers, to sell plundered knickknacks. With the proceeds she and Mary would buy new clothes for themselves, silk scarves and stockings, and theatrical Russian cigarettes, black with gold tips. To reward the younger children for running these illicit errands, Kathleen sometimes took them to a matinée at the new cinema in Walsall. They would sit in the dress circle and be served tea and toasted tea cakes during the interval, while Kathleen smoked, acting the part of Lady Bountiful to perfection. She and Mary also bought books. When Dr. Walter caught them reading *Madame Bovary*, he summoned all the children to the nursery and burned the book in front of them on the fire.

The strict system of chaperonage was only just coming to an end. Kathleen and Mary caused something of a scandal locally when they ordered a drink at a local miners' pub, something wholly out of place for unaccompanied women of their age and social standing. Such a gesture was typical of the Garman siblings: flying in the face of tradition, headstrong, oblivious to criticism. They had the confidence that is often a hallmark of

Kathleen, circa 1920.

people born into a large family, and they could not have been unaware of the power exerted by their remarkable appearance. All of them were handsome at the very least, and some of the family had film-star good looks. Like Kathleen, Mary was beautiful, with an ethereal quality which hinted at the religious fervor she would develop in later life. The whole family were tall, and the girls had long, thick hair and dramatic arched brows. But their eyes were most distinctive. Blue, violet, or dark, they all had eyes that were wide, expressive, and velvety. Everybody noticed them.

The war created a prevailing sense of gloom and menace. The children were made to knit khaki scarves for the soldiers, and Mrs. Fowl, the housekeeper, insisted on taking the children down to the grimy railway station to cheer the departing troops and, later, the Red Cross trains bringing the wounded men home. None of Garmans' immediate family had been affected, but the newspaper's long black columns listing the men killed in battle were sad enough. Three-quarters of a million British men were killed in the war, and another two hundred thousand were left disabled. All the suitable young men in Wednesbury had gone off to fight in the First World War, leaving only dull curates as potential suitors. Then there were the ravages of the subsequent influenza epidemic, which meant that between the two world wars there were far more women than men in the country. Walter would have liked his daughters to have married into the clergy, but they had other ideas. They knew he would not give his permission for them to leave home, so Mary and Kathleen simply packed their things and ran away.

Chapter Two

LONDON

Twenty-year-old Mary and eighteen-year-old Kathleen arrived in London in 1919, penniless. Angered by their boldness, their father refused to give them any money to live on, so Kathleen took a job helping with the horses that pulled the Harrods carriage and also posed as an artist's model. Daringly for a middle-class young woman of the time, Mary drove a delivery van for Lyon's Corner House. Eventually, Dr. Garman relented and gave both of his daughters an allowance, which enabled them to give up their jobs and enroll at a private art school called the Heatherley.

The two Garman sisters soon found themselves caught up with the artistic and musical set which gathered at the Café Royal. These mirrored halls became legendary as the crucible for bohemian London. The smoke of Turkish tobacco rose to the painted ceilings while people conversed on crimson velvet banquettes below. Oscar Wilde and Max Beerbohm were

among the first to make the place their own. Verlaine and Rimbaud had also visited, and when Rodin came to England he had
been given a special dinner there. Everyone drank, some to
excess. A Henry Bateman cartoon showed the whole room roaring with laughter at a wide-eyed newcomer, with the caption:
"The Girl Who Asked for a Glass of Milk at the Café Royal."
Walter Sickert and the Sitwells and the "Bloomsberries"—
Virginia Woolf, E. M. Forster, Maynard Keynes, Lytton Strachey,
Roger Fry, and others—made it their own. By the time Mary
and Kathleen came to London, Wyndham Lewis, D. H.
Lawrence, Jacob Epstein, Nina Hamnett, and Aleister Crowley
were regulars. Augustus John more or less lived there.

Like the rest of the Café Royal habitués, the Garmans fre-
quented two restaurants which were to become crucial in their
lives. The Eiffel Tower at 1 Percy Street and the Harlequin at
50 Beak Street were the prototypes for the kind of selective
hospitality and inebriation that would later solidify into Soho
legend, in places like the Colony Room and the French Pub of
the 1940s and '50s. The Eiffel Tower was run by Rudolf Stulik,
an Austrian restaurateur of the old school, and a character. He
entertained royalty but also befriended many penniless artists
whom he would feed, clothe, and shelter at his own expense. He
recouped some of his losses by charging outrageous prices and
ignoring licensing laws: customers were allowed to stay as long
as they liked, often well into the next morning. On these occasions someone would invariably ask Mr. Stulik to join the party
and partake of a glass of wine, to which he always gave the same
reply: "I *never* drink with my customers"—long pause—"but

tonight I shall make an exception." A private upstairs dining room had been decorated by Wyndham Lewis in 1915. He often dined there, as did Lady Diana Manners (later Cooper), the Asquiths, Maurice Baring, Roy Campbell, Tallulah Bankhead, and the various members of the Bloomsbury contingent. Augustus John had his own table in a far corner, kept permanently reserved for his exclusive use. Michael Arlen, author of the much celebrated *The Green Hat*, was a regular, as was Nancy Cunard. The Harlequin was decorated with painted panels, the work of Alvaro de Guevara (who painted a famous portrait of Edith Sitwell), assisted by Roy Campbell. Like Stulik, the proprietor gave credit to regulars, which meant he experienced occasional difficulties in making ends meet. Roy Campbell would sometimes hand his whole paycheck over to the owner, instructing him to deduct from it daily the cost of his food and drink.

Various members of this crowd began to press their attentions upon Mary and Kathleen. Through her regular attendance at concerts, Mary had attracted the notice of Bernard van Dieren, the Dutch composer. Being middle-aged and married did not stand in the way of his ardor, and after he had courted her for almost a year she at last invited him home. Once van Dieren became a regular visitor, he prepared delicious meals for the girls on their gas stove, which delighted them since they still had no idea how to cook. He dedicated songs to Mary and sent her letters beginning, "Dolcissima, carissima Maria." She would sit at the battered piano Kathleen had installed and attempt to play his highly complex and all but illegible scores, while the composer stood behind her, wincing at her mistakes but smitten. In later years, Mary would make her husband jealous by naming a kitten Bernard after her admirer.

The sisters lived in a one-room studio on the ground floor of 13 Regent Square, at the very edge of Bloomsbury. The place was lit by naked gas flames, which were picturesque but dangerous. Shortage of money and a reaction against the dark, Victorian appointments of Oakeswell were largely responsible for the sisters' particular style, which involved unframed drawings and scented geraniums and simple divans, which doubled as beds and as sofas. Such underfurnishing was very bohemian, against the prevailing brocade and pleats and pelmets of the time. The Garmans found that a few flowers in a tall vase and a fire in the grate could make a place look wonderful.

The artist Bernard Meninsky, who was teaching at Westminster School of Art, became another suitor of Mary's, painting her portrait more than once. Ferruccio Busoni, a composer who boasted that he had heard Brahms play and that he had himself performed for Liszt, also pursued the sisters. In his memoir *Broken Record*, Roy Campbell recalled that there had been dozens of other admirers around Mary and Kathleen in those days, listing "highbrows, Jews, poets, authoresses, painters, pianists, singers, ballet dancers, and even an economist. No other contemporary women ever had so much poetry, good, bad and indifferent, written about them, or had so many portraits and busts made of them."

After seeing Tamara Karsavina and Nijinsky dance for the

London season of Diaghilev's Ballets Russes, the sisters began to adopt a Russian style, swishing about in long cloaks and fur hats and high boots. A photograph shows Kathleen in side-buttoned shoes, white stockings, and an extravagantly fringed white shawl, wearing a turban on her head, pinned with an enormous brooch. In warmer weather they were among the first respectable women in London to go hatless, something only prostitutes had done until then.

One day in August 1921, Kathleen and Mary were having dinner at the Harlequin when Kathleen became aware of a strange man who kept staring at her from the other side of the room. He was unkempt, with wild hair, but the intensity with which he regarded her, the concentration in his blue eyes, made her unable to look away. Then a waiter brought over a note: would the women join his table? Kathleen was amused, but she and Mary left without accepting the invitation. It was a different story when she found herself back in the Harlequin, alone, a few days later. The man was there again, by himself. He sent another invitation to her, and this time she joined him. He told her he was a sculptor, Jacob Esptein, and at once asked her to sit for him.

Kathleen did not know him by sight, but she would surely have heard of him by reputation. At forty, he was a controversial artist, and a highly successful one. Epstein had come to Europe from his native New York, living for a time in Paris before settling in England. He had taken British nationality in 1910, on the advice of George Bernard Shaw. Augustus John had befriended him. At his first major show at the Leicester Galleries in 1917, Epstein had sold sixteen hundred pounds worth of bronzes. His portrait busts were, even then, a lucrative addition to the carved work which always brought him so much publicity, even notoriety. He had already made the tomb for Oscar Wilde in the Père-Lachaise Cemetery in Paris. Possibly the only piece of sculpture ever to have been placed under arrest, it was kept hidden by a tarpaulin, guarded by police, from its installation in 1912 until the outbreak of the First World War. The naked figures that he made for the British Medical Association building in the Strand (the building still stands, on the corner of Agar Street) had caused an uproar in London. At the initiative of Augustus John, a senior bishop had been called in to declare that they were not, in fact, indecent. His statue *The Risen Christ*, exhibited at the Leicester Galleries in 1920, had caused further outcry. The writer John Galsworthy was observed coming away from viewing it with a red face, fists clenched: "I can never forgive Mr. Epstein for his representation of Our Lord," another visitor told the gallery's proprietor. "It's so un-English."* Several of the models Epstein had been in the habit of working with were women of dubious reputation, and his use of unclothed models was the stuff of gossip. He was also a married man and the father of a young child.

Epstein had spent a great deal of time studying Egyptian art at the British Museum (just around the corner from his home in Guilford Street, Bloomsbury), sometimes taking the young

---

* *The Risen Christ* was bought in 1920 for twenty-one hundred pounds by the polar explorer Apsley Cherry Garrard, author of *The Worst Journey in the World.* He kept the figure in his garden.

painter Mark Gertler with him. That Kathleen looked rather Egyptian may have accounted for some of her fascination to Epstein, who was always attracted to the exotic. She wore her hair straight, with long, straight bangs, in the style of Cleopatra, and she had prominent cheekbones. Later, Epstein gave her an ancient mask, said to be the face of Nefertiti, which looked very much like her. He called Nefertiti the "calm-faced Queen, with her cold, mysterious glance." Kathleen's sultry appearance was at variance with her unruffled temperament, and this air of tranquility and self-containment made her beauty all the more remarkable. Epstein was very striking, too, with his extraordinary energy and soft, New York–accented voice. The attraction between them was instant. The morning after their first night together, he began to make a first sculpture of her head.

Kathleen, at around the time she met Epstein in London, 1921.

Margaret Epstein had always been tolerant of her husband's affairs, encouraging his models and mistresses to come to live with them, preferring to keep her enemies close. The extent to which she was prepared to accommodate her husband's infidelity is evident from the fact that she was bringing up Peggy Jean, his daughter by a former lover, the beautiful Meum Stewart. But Kathleen, nearly thirty years Margaret's junior, was different. She was never just another of Epstein's infatuations, whom Mrs. Epstein could brush off after he'd finished sculpting her. She wasn't biddable. More than a lover, she was a parallel wife. From the beginning, Mrs. Epstein disliked her intensely, rightly intuiting that she was to be her greatest rival. Little Peggy Jean was told that Kathleen was a witch, and dangerous. Her earliest memory was of hiding behind her adoptive mother's skirts as Kathleen was approaching along the other side of the road in her billowing black cloak.

Only two months after Kathleen met Epstein, her sister met the man that she was to marry. Mary had seen Roy Campbell on a bus and was so taken with him that she had jumped off and followed him to the Eiffel Tower restaurant. But when she saw Iris Tree, with whom he was entirely innocently lunching, she did not dare go into the restaurant. A fortnight afterward, fate intervened. Roy's rakish friend Stuart Gray had been evicted from his lodgings having, according to Campbell, annoyed his titled landlady by making a pass at her cook. The Garman sisters took pity on him and said he could take refuge in the coal cellar beneath their house. Gray duly asked Campbell to take some of his belongings to Regent Square in a handcart. "When I had trundled this cart round to Stuart Gray's new quarters, the most beautiful young woman I have ever seen opened the door to me," Campbell later recalled. "When I saw her I experienced, for one of the few times in my life, the electric thrill of falling in love at first sight." He was just twenty; she was twenty-three. South African by birth, he had studied at Oxford before coming to London.

That a future husband should simply turn up on the doorstep was typical of the Garmans.* "I was only too glad to meet my wife, and get married within a couple of days of seeing her," Campbell wrote in *Broken Record.* This was a slight exaggeration, since the wedding did not actually take place for another three or four months. But within three days he had moved into the girls' studio room. Tall and thin, with startlingly blue eyes, he was already writing poetry, living on beer, and forgetting to eat, or eating only radishes, leaves and all, bought from a market stall. The girls decided to fatten him up, and the three of them would lie, arm in arm, in front of the fire while he read them fragments from the poems that would become his first book, *The Flaming Terrapin.* Rumors of a ménage à trois soon began to circulate, but from their childhood the Garmans were well accustomed to living cheek by jowl with members of both sexes, and there was no substance to the gossip that both girls had become Campbell's lovers. Epstein was always extremely jealous about Kathleen, in spite of his own unconventional domestic arrangements. On one occasion he came to Regent Square and launched into a tirade against Campbell, warning them to beware of him. Meanwhile, unbeknownst to him, Campbell was hiding behind the piano, popping up to make silly faces at the sisters, who were both struggling not to laugh.

* Many years later, an English nephew of Mary's would happen upon the French woman who would become his wife, in a revolving door in Texas.

In December 1921, Roy was brought home to Oakeswell to meet Dr. and Mrs. Garman. Perhaps the future bride took advantage of Kathleen's favor with Dr. Walter, for it was she who introduced him: "Father, this is Roy, who's going to marry Mary." The doctor was less than delighted: his prospective son-in-law was penniless, a stranger to him, and, he observed, a dipsomaniac. Campbell had tried to conceal his drinking by abstaining during family meals, but he had taken the fifteen-year-old Helen with him to a local pub, as well as begging ten-year-old Lorna to fetch him some beer to help him get rid of a hangover. In fact, Roy's background was not so very different from Mary's, since his father was a widely respected doctor who also treated the poor free of charge. It must also have mitigated in Roy's favor when Dr. Walter discovered that one of his own sisters had been treated by Dr. Campbell while living in South Africa.

After Christmas at Oakeswell, Roy and Mary went to stay with Augustus John at Alderney Manor, outside Parkstone in Dorset. Here they enacted a sort of gypsy wedding, with Trelawney Reed as the priest and John—who christened Mary "Little Lord Fondle-Roy"—as acolyte. A gypsy wedding ring was procured, and after the ceremony, the lovers were ceremoniously conducted to a cottage nearby and put to bed. "It was good fun," wrote John, "and I thought wanted very little to make it a true and valid ceremony. As a matter of fact, there could have existed no closer union in the world than that of Roy and Mary Campbell."

Two months later, their real marriage took place. Roy had

borrowed an ancient frock coat from a waiter at the Eiffel Tower and flattened down his hair, so that he looked like an undertaker. When he arrived at Wednesbury, Mary almost fainted with horror. She begged him to put on his old suit, which he did. But when he knelt at the altar, the smarter guests noticed that he had holes in the back of his shoes, padded with newspaper. Mary's attire was also unusual. She wore a long black dress and a golden veil, apparently because she had no money to buy anything else. It is hard to believe that Dr. Garman would not have provided a wedding dress for his eldest daughter, had she asked him for one, and it seems more likely that her dress was intended to *épater les bourgeois*. The oddness of the pair's dress made the Garmans' old nanny, Ada, wail, "I always thought Miss Mary would marry a gentleman, with a park!" After a wedding breakfast at Oakeswell, the couple returned to London and a party at the Harlequin before leaving the celebrations to retire upstairs to bed. Wyndham Lewis described the farcical drunkenness of the occasion:

> The marriage feast was a distinguished gathering, if you are prepared to admit distinction to the Bohemian, for it was almost gipsy in its freedom from the conventional restraints... Jacob Kramer and Augustus John were neighbours at table and I noticed that they were bickering. Kramer was a gigantic Polish Jew and he was showing John his left bicep... John did not seem interested. But Kramer would not be put off or have his bicep high-hatted like that. It went on moving about under his sleeve in an alarming fashion, and Kramer looked at it as if ir-

resistibly attracted. It was such a funny thing to have in your sleeve, a sort of symbol, he seemed to feel, of Power. He tapped it as a prosperous person taps his pocket.

"I'm just as strong as you are John!" he kept vociferating, screwing his neck round till his nose stuck in John's face.

"You've said that before," John answered gruffly.

"Why should I put up with your rudeness, John—*why*! Tell me that, John! You're a clever man. Why should I?"

John shrugged his shoulders and looked down rather huffily at his spoons. He sought to indicate to his neighbour that philosophically interesting as the question might be, it was no time to discuss it, when we were convivially assembled to celebrate the marriage of a mutual friend.

At this moment Roy Campbell entered in his pajamas. There was a horrid hush. Someone had slipped out to acquaint Campbell with the fact of this threat to the peace. In a dead silence the bridegroom, with catlike steps, approached the back of Kramer's chair. That gentleman screwed round, his bicep still held up for examination. He was a very supple giant, and by this time he was sitting one way but facing another, having as it were followed the bicep round behind himself.

"What's this, Kramer?" barked Roy, fierce and thick, in his best back-veldt. "What are you doing, Kramer?" Roy Campbell pointed his hand at his guest and began wagging it about in a suggestive way as if he might box his ears or chop him on the neck with it.

"Nothing, Roy! I'm not doing anything, Roy!" the guest answered, in a tone of surprise and injured innocence.

"Well, you let John alone, Kramer! Do you hear!"

"I'm doing nothing to John, Roy. I was talking about painting," Kramer said.

"Never mind painting, Jacob. Is that how you talk about painting, Jacob?"

"Yes, Roy," said Kramer, in an eager and conciliatory voice. "I get worked up when I talk about painting, Roy."

"Look. Could I throw you out of that window if I wanted to Jacob?"

"Yes, Roy, you could," said Kramer humbly. "You could, Roy."

"You know I could, Jacob?"

"I know you could, Roy." Kramer nodded his head, his eyes screwed up.

"Well then let my guests alone, Jacob. You let my guests alone. Don't let me hear you've interfered with John again. Mind I'm only just upstairs, Jacob. I'll come down on you!"

A strangled protest and assent at once came from Kramer; and stiffly and slowly, his shoulders drawn up, his head thrust out, in apache bellicosity, Campbell withdrew, all of us completely silent. When the door had closed, Kramer got up, came round the table and sat down at my side. He'd put his biceps away. He continued with me the conversation about painting which had taken such a personal turn on the other side of the marriage board.

The Campbells rented the top floor of 50 Beak Street, above the Harlequin. Mrs. Epstein (or possibly her jealous husband himself) was thought to have set the waiters there to spy on Kathleen and Mary, in the belief that Kathleen might be caught in some indiscretion. Epstein still half-believed that Campbell was bedding both sisters, while Campbell thought he was an absurd figure and had nicknamed him Sennacheribs. One evening, Roy and Mary were dining with Augustus John, when Epstein passed their table, glowering. Sending his wife out with John, Campbell invited Epstein to "have a word" upstairs. The restaurant fell silent as the two men mounted the stairs. "I backed into the room and as he followed dodged round him and locked the door," wrote Campbell, "he tried a couple of dirty kicks: then he got very frightened indeed and started yelling as I advanced...a chest of drawers fell over. I sat quietly on his stomach as he lay philosophically blinking at the ceiling and quite conscious. The whole house had been shaken and there was a crack in the wall...I went out to find Augustus John and Mary, having sustained nothing more than a scratch on the forehead from Sennacheribs' waist-coat buttons as I threw him over my head."

After Mary's marriage, she and Kathleen were never as close as they had been when they shared the studio in Regent Square. The mutual disdain of Campbell and Epstein must have played its part in their estrangement, even though it never erupted into violence again. Some of Campbell's views would have been anathema to his future brother-in-law and to Kathleen. His much-vaunted right wingery was expressed with a brazenness which was more provocative than serious, but he succeeded in annoying many. Epstein had frequent occasion to defend himself against anti-Semitism, not least when the Christian images he made outraged the more reactionary spectators. Certain of his public sculptures were vandalized repeatedly, and at the end of the 1920s, his house was daubed with swastikas and scrawled messages like "Jew go home." In his memoirs, Campbell would write, "I am no pogromite myself. One can forgive the Jews anything for the beauty of their women, which makes up for the ugliness of their men. But I fail to see how a man like Hitler makes any 'mistake' in expelling a race that is intellectually subversive as far as we are concerned: that has none of our visual sense, but a wonderful dim-sighted instinct for dissolving, softening, undermining, and vulgarising." No wonder there was enmity between Epstein and Campbell.

Campbell, too, had furious bouts of jealousy. He described Mary as "the only woman with whom I have found that jealousy or quarrels acted mutually as a kind of aphrodisiac." To this end, she may have provoked him deliberately. In his memoirs he refers more than once to the fact that his wife was a known "man-hater," a veiled way of saying lesbianism. "That she was said to be a man-hater never daunted me for a moment, but only made it more of a challenge, more exciting, and more of a responsibility to capture her for myself for ever." Mary's youthful infatuation with a woman art teacher at school had already been rumored. Whether it was to this distant romance that she was referring or to some more recent attachment, Mary caused a row not long after her wedding by stretching out her arms and murmuring, "Oh, how lovely she was!" in *Light on a Dark Horse*, Campbell recalled the incident, although, tellingly, he omits the remark that occasioned the quarrel:

Though we were very happy, my wife and I had some quarrels since my ideas of marriage are old-fashioned about wifely
obedience and in many ways she regarded me as a mere child
because of being hardly out of my teens. But any marriage in
which a woman wears the pants is an unseemly farce. To shake
up her illusions I hung her out of the fourth-floor window of
our room so that she should get some respect for me. This
worked wonders for she gazed, head-downwards, up at the stars
till the police from their HQ on the opposite side of Beak
Street started yelling at me to pull her back. She had not
uttered a single word and when I shouted out pleasantly across
the street: "We are only practising our act, aren't we, Kid?" she
replied "Yes," as calmly and happily as if we did it every ten
minutes. The police then left us alone, saying: "Well, don't prac-
tise it so high over other people's heads, please."

My wife was very proud of me after I had hung her out of
the window and boasted of it to her girl friends.

Among these girl friends was a young ballet dancer called
Jeanne Hewitt. She and Mary had met while attending the
same art school and at once found each other congenial. Like
Mary, Jeanne was one of a large family who shared the
Garmans' interest in music and the arts. Several of the Garmans
and Hewitts got to know one another during the early 1920s
and were on friendly terms. Kathleen gave piano lessons to
some of the younger Hewitts. Later, when Mary brought
Jeanne home to stay with her family, she told her, "You and my
brother Douglas will fall in love." And they did.

Having survived for a time by pawning their wedding presents, the Campbells ran out of money, and in 1922, they
decided to leave London. Mary proposed that they should live
in Wales, at Aberdaron, the farthest point of Gwynedd. Here
they could live cheaply and without the distractions of Soho
and Bloomsbury, which included heavy drinking. It was a place
that Mary knew from childhood holidays, when a house would
be rented for the summer and the strict discipline of home
would be relaxed, with the children free to swim and explore
and wander to their hearts' content. The Campbells took a converted stable called Ty Corn, about three miles out of the village. The rent was thirty-six shillings a year. They found that
they were able to live for very little, thanks in large part to the
generosity of the locals who, in time-honored Welsh custom,
welcomed their arrival with gifts of sacks of potatoes, peat, and
hens. Campbell hunted (or poached) snipe and rabbits and
pheasants. Ty Corn "leaked every way, had only a mud floor,
and the wind whistled through the walls, but we had the time
of our lives there living on the continual intoxication of poetry
for two years," recalled Campbell. The couple read aloud to
each other, consuming the whole of Milton, Marlowe, Shakespeare, Dryden, Pope, and many of the Elizabethans. Dr.
Campbell, who had stopped his son's allowance in protest at his
hasty marriage, became reconciled and resumed the ten pounds
a month, most of which Mary and Roy spent on books.

Augustus John's son Romilly ran away from home and turned
up at Ty Corn, where he stayed for a few weeks. In his autobiography, he remembered how depressed he had been by the

humbleness of the abode, but concluded, "I really think it never crossed his [Campbell's] mind that the cottage in any way fell short of a desirable country residence; nor did it, when the holes had been stuffed up and a great fire was roaring up the chimney." The beer which Campbell would carry back from the village in a gallon jar contributed to their good spirits. Years later, Mary said that she and Campbell had been the first hippies. They attracted the notice of the locals by wearing flowing and brightly colored clothes, and Campbell's hair was long, by the standards of the day. They were known for making love on the clifftops in broad daylight. The walls of their cottage were decorated with charcoal sketches of each other naked (like his wife, Campbell was very good at drawing). When the coalman came by one day, Mary opened the door, with one of her sisters standing behind her, both of them in the nude. The sight so distracted him that he forgot what he was there for.

In September 1922, Campbell finished the long poem which would become *The Flaming Terrapin*. He made five copies and sent them to his closest friends for comment. The first to respond was the critic and radical Edgell Rickword, at that time himself a poet of some renown. Campbell was euphoric at his letter ("I simply fell down on my bed and howled like a baby when I got it," he told his mother), which said: "the poem is magnificent... I know of *no-one living* who could write in such a sustained and intense poetical manner... Good luck and ten thousand thanks for such a poem." Augustus John showed the poem to T. E. Lawrence, who in turn wrote to the publisher Jonathan Cape, urging him to "get it—it's great stuff."

There was further cause for celebration when, almost nine months to the day after their wedding, the Campbells' first daughter, Tess, was born in the middle of a violent November storm. "I have not seen anything to equal the extraordinary courage of my wife in fighting through this fearful night, when the wind blew the tiles off our roof and the rain and wind rushed in headlong," wrote Campbell. A young Welsh midwife delivered the baby by the light of a single oil lamp that swung wildly from the rafters in the rising wind. Campbell went and sheltered behind a piece of corrugated iron on the beach, suffering sympathetic pain. At dawn, when the storm calmed, he went out and shot a snipe and grilled it on a spit to bring in on a tray for Mary's breakfast.

Epstein, meanwhile, came to see Kathleen every evening after work, often bringing gifts. On Wednesdays and Saturdays he stayed the night with her at Regent Square, returning home through the empty streets at dawn. Whenever they were apart, he wrote her passionate letters. Jealousy continued to trouble him.

I wrote you yesterday hurriedly and feeling somewhat wretched. This naturally the result of hearing some gossip about you. Then I could hear nothing of what you said on that miserable Café Royal phone in the evening. Why do we meet to part and yet it is necessary. Some day it may not be. How well you know me. What you write is true and from the moment I meet you my elation is equal to yours and my happiness so genuine that you can easily discern it . . . Write me:

your letters console me and dissipate the things I hear and often imagine to be true. What a lack of faith in myself to be thrown from the heights to which you raise me and realise what a mockery our love is. My hell is almost as intense as the heaven I've been in. When you say they are liars I believe you. I want your love and without it I would be the unhappiest man on earth.

He had made his first head of Kathleen immediately after they met. A second, which he began working on in the early months of 1922, depicted her as a woman sated with love, lips parted, eyes half closed, rapt. If his wife saw this head in his studio, and since she managed all his affairs it is more than likely that she did, her worst fears about her rival were surely confirmed by its tender sensuality. Around this time Epstein wrote to Kathleen,

I have your note and the news that you arrived home. I love to read what you feel and that you know my happiness was so extreme that were I to die then and there I would be an image of supreme happiness in death. I still live in the nights, the 3 nights we were together. I express myself readily enough with you but to write of it is impossible in me, our nearness, your glances and every touch of you fills me with joy & that is what makes it difficult for us to be together for long. I will try now and finish the head I am on: even unfinished I think it will be beautiful, perhaps owing to the subject. The true artist should make anything wonderful. When I look on you and am with you I see the most wonderful things and they are to be but a memory with me. No more. Write and tell me how you are.

In the spring of 1923, Kathleen paid a visit to Oakeswell. Epstein wrote to her there,

> It was terribly trying to hear you on the phone tonight. You asked me if I was happy. Are you? I am working and looking forward to you. I tried to tell you that if I dreamt of you as I do, it was only a suggestion of the reality which is the best thing I know. The last letter I wrote you was on a foggy day, last Tuesday. I went out filled with the idea of you and imagined our waking up on such a day and lying all through it together. I walked the streets and thought of you... You are my sweetheart having my thoughts by day, my dreams, my ardent longings by night... it is real torture to think of you. You only can know what I mean by this.

He asks whether Kathleen is happy. It seems likely that she should have been, surrounded by her family and the familiarity of Oakeswell, with her birthday to look forward to. It was also a place where she could indulge in her favorite activity: reading. She loved Dostoyevsky and Goethe, Flaubert and Thomas Hardy, and Dickens. Henry James was her favorite. She also adored poetry and was to number some of its most distinguished practitioners among her friends: Theodore Roethke, T. S. Eliot, and Allen Tate among them. While she was at Oakeswell, she read aloud to Dr. Walter, who had been suffering from heart disease. When he died, Kathleen was at his side. The family believed that he had been killed by overwork, for he was only sixty-two. Certainly, his dedication to his patients was such that many hundreds lined the streets to pay their respects as his cortege passed by. The family was plunged into mourning. All, that is, except the spirited little Lorna. Only twelve years old at the time, she was told of her father's death while she was away at boarding school. As soon as she heard the news, she jumped over the tennis net for joy, knowing this meant that she would be able to leave the hated school and begin a new life of freedom.

Chapter Three

Sunsets etc

Lorna was to tell her children that the family had been so poor after Walter's death that they had had to tear up the carpets to sleep beneath, because of the awful cold. They were so hard up, she said, that she had only a school gym tunic to wear, which became shorter and shorter with each passing year. Contemporary visitors recall that the young Garman girls actually wore pale chiffon dresses for outings in the evening and perfectly ordinary wool and tweeds for other occasions, although a photograph of the time does depict Lorna in a tunic, wearing shoes with neither stockings nor socks. It was true that Mrs. Garman and her companion, Miss Thomas, bickered politely about whether they could afford to put another piece of coal on the fire, and the youngest three children—Mavin, Ruth, and Lorna—were all obliged to leave their schools. There was not enough money for Mavin to attend the university.

Dr. Garman had not owned Oakeswell. Renting a house,

even such a large house, was common at the time, so there was nothing unusual in his not having property to bequeath. He had plenty of possessions: furniture, books, pictures, prints, silver-plated articles, linen, china, and glass were all left to his widow. The more valuable wines and liquors and motorcars (to have owned more than one must have been rare in the early 1920s and suggests he may have been extravagant) were sold and their value added to the estate. Small bequests were left to specific children, but Kathleen received nothing. Perhaps her father wanted to punish her for the affair with Epstein, which was said to have dismayed him greatly.

The doctor's will was published in the *Times* on October 19, 1923: "Garman, Dr Walter Chancellor JP of Wednesbury, Staffs, for nearly forty years Medical Officer of Health for the town . . . £24,151." Of this, £21,076 remained, after tax and legal fees[*] a large amount of money, which should have amply supported his wife and children. Yet the signs are that it did not. The money was in shares of the Steel Nut and Joseph Hampton Ltd, a local firm in which various members of the Garman family had an interest. Steel prices had slumped after the First World War, and it seems that Dr. Walter's investment did not prosper.

Mrs. Garman did not—or could not—retain staff, although she and her children were to remain in contact with their former nanny, Ada Newbould. Only one person stayed with them, the children's governess, Miss Elizabeth Thomas, whom everyone called Tony. A Victorian figure always dressed in rusty

---

[*] A sum of the order of about £630,000 in early twenty-first-century value.

black, she was tireless, uncomplaining, fiercely loyal—she stayed with the family until she died—and occasionally stern, more a figure of respect than of love. Mrs. Garman was vague and sweet and smiling but impractical. Sometimes she would hold a duster, looking bewildered. In her prayers she implored the Good Lord to spare her the drudgery of "abhorrent" house-work. Her daughters mostly inherited her dread of housework: Kathleen's washing up was smeary, and there was a thick accu-mulation of dust in Lorna's house. Tony Thomas seems to have been the answer to Mrs. Garman's prayer. It was she who kept the household together, assisted most mornings by a young maid of fourteen or fifteen, who would come in from the vil-lage. Miss Thomas taught herself to cook from Mrs. Beeton and would go out to the shops to order food and pay the bills. She made soda bread and apple dumplings. She nursed any children who were sick and taught them English and rudimen-tary French and how to write script. She laid the fires.

Oakeswell was too big for Mrs. Garman and Miss Thomas and the three children who were still at home. In 1924, the dwindling number of Garmans moved to Black Hall, an ancient half-timbered farmhouse at King's Pyon in remote Hereford-shire, rented for sixty-two pounds per year. It was cheap, with plenty of room for the three youngest, and Mrs. Garman pre-ferred the country, with its broad hedgerows and orchards, to the industrial Midlands. Behind the house were barns and a yard, and at the front an area of lawn before the ground sloped away toward a large pond fringed with trees. When Lorna and Ruth were in their teens, they used to take the gramophone out onto the grass and practice dancing the Charleston in their bare feet. The place was barely a hamlet, just a handful of houses and a church surrounded by low hills and fields of pasture reached by narrow lanes, but even in the remote Welsh marches, jazz was in the air. The Garmans became mad about dancing and were soon attending village hops for miles around. Ruth and Lorna with their looks and reckless gaiety were besieged by boyfriends, although both girls were still only in their early teens. They would be collected by young men in long motorcars and driven off into the night to dance. Their brother Mavin would go dancing with them, having found a local job on a farm. Years later, he remembered the intoxication of those country evenings, when they drank cider and forbidden swigs of whisky, and all the young Herefordshire women wore a scent called Ashes of Roses.

Both the girls were extremely precocious. They used to make wreaths of wildflowers and hang them on the war memorial in the churchyard, parodying the words "They fell for us" into the heartless "They fell for US." The young Garmans were certain that they were different and better than those around them. In these remote surroundings, the three relied on one another for company, and they became unusually close, sharing everything and often sleeping in the same room, even the same bed. Mrs. Garman later admitted to Lorna that their closeness had made her more than a little apprehensive. In adult life, their friends sometimes wondered whether there had been an incestuous bond among them, but Mavin maintained that there had not, although it was not scruples or taboo which stopped it, but that there were three of them, which prevented pairing.

Sylvia and the young Lorna at Black Hall in Herefordshire, 1924.

It was Mrs. Garman who urged Mary and Roy Campbell to take the infant Tess away from Wales. She felt that the extreme rusticity of their house was unsuitable for bringing up a baby, so later that spring they moved back to London. *The Flaming Terrapin* was to be published in May 1924, and the flurry of literary interest Roy received, even before the book came out, made him eager to be in the city once again. They went back to Fitzrovia (an area just north of Soho, where artists and writers convened at the Fitzroy Tavern), taking a room at 90 Charlotte Street. Campbell started going to rackety pubs and boxing matches, where he drank whiskey with the wild and inebriate artist Nina Hamnett, to his wife's disapproval. But he did help with the baby, walking her about strapped to his back in a blanket, Zulu style, to quiet her.

Nina Hamnett lived in nearby Fitzroy Street. Walter Sickert was also a neighbor, often to be seen emerging from the basement of his studio in eccentric dress: sometimes he would appear in a tweed deerstalker hat with earflaps and a poacher's jacket, sometimes in a bowler hat and city suit. A young artist named Kathleen Hale (later to be famous as the creator of the children's books featuring Orlando, the marmalade cat) had moved into a room in Fitzroy Street, very near to the Campbells. Their circles overlapped: Hale did occasional secretarial work for Augustus John and was friendly with Epstein, too. Vanessa Bell, Duncan Grant, Jacob Kramer, and Bernard Meninsky were all nearby. A few paces away in Fitzroy Square, Roger Fry had founded the Omega Workshops to give employment to promising young artists, decorating and designing pottery, furniture, and textiles. McKnight Kauffer, who designed the distinctive 1920s advertisements for the Underground, was another tenant of the square.

The area was full of merrymaking and the sense of freedom. Even on Sundays it had none of the Sabbath gloom that prevailed in those years. The inhabitants of Bloomsbury and Fitzrovia flung out the rule book and lived as they pleased. Everyone stayed in bed until noon, when the men would emerge in their shirtsleeves with empty jugs to fill with beer from the local pubs. A Dickensian stream of street peddlers passed through. The muffin man would arrive at teatime, ringing a handbell and balancing a wide tray of muffins and crumpets, covered with a clean white cloth, on his head. There was the cat's-meat man who toured the street, followed like a pied piper by a meowing, undulating stream of multicolored cats. Everyone loved the organ grinder and threw pennies to him from their windows. He attracted hordes of children, who skipped and danced and turned cartwheels to the wheezy and raucous melodies in the safety of the empty street. A troupe of fantastic young men with their own barrel organ used to arrive, prancing and pirouetting along Fitzroy Street, wearing thick makeup, feather boas, and huge battered picture hats, with their trousers showing below bedraggled skirts. There was the throb of rehearsing dance bands, and the piercing whine of the knife grinder's machine as he turned the grindstone with a pedal worked with one foot. Out of school hours, the street would ring with the shrieks and squalling of ragged children who swung on pieces of rope tied to lampposts and who played at hopscotch

and marbles. The Café Royal crowd would lurch in and out of
local pubs and restaurants. This was the London that Louis
MacNeice described: "foreign names over winking doors, a
kaleidoscope / Of wine and ice, of eyes and emeralds."

The Flaming Terrapin, published simultaneously in England
and America, was very well received, but the Campbells were
still hard up. When Roy Campbell's cousin visited Charlotte
Street, she was taken aback to see how hungry and poor they
were, and straightaway took them to a nearby restaurant. Anxious about being able to provide for Mary and Tess, Campbell
turned, as ever, to drink: the more he fretted, the more he
drank, and the more he drank, the less money he had. One
snowy March evening, Mary came back to their room to find
that her husband had locked the door and gone to the pub. She
sat for hours on the doorstep in the bitter cold, comforting
the crying baby until a drunk and penitent Roy returned. For
the first time since her wedding, she wondered if her father had
been right to disapprove of the marriage.

It was a difficult time for Kathleen, too. Epstein's letters suggest that he was constantly suspicious of Kathleen seeing
other men, an idea that his wife was only too keen to foster. In
addition to bribing waiters in the hope of catching Kathleen in
some indiscretion, the desperate Mrs. Epstein may have hired a
private detective. Kathleen often had the sensation that she was
being followed. Whether the spy was set by Epstein or by his

wife, a friend remembered seeing a man lurking in Regent Square, watching Kathleen. Eventually she grew used to this man's presence and enjoyed playing cat and mouse with him.

In the summer of 1923, Mrs. Epstein's jealousy of her young rival erupted. Late one night, a note came through the door in Margaret Epstein's handwriting, summoning Kathleen to a rendezvous. Kathleen was never a coward, either physically or morally, so she went to confront her at Guilford Street. Epstein seems not to have been at home. The frantic Mrs. Epstein took Kathleen into a room and locked the door, before producing a pearl-handled pistol from under her capacious skirts. Then she said, "I'm going to kill you," and shot her. The bullet hit Kathleen just to the right of her left shoulder blade, whereupon Mrs. Epstein panicked and ran out of the room, leaving the bloodied Kathleen to stagger out into the street alone. Her brother Douglas was outside, and she collapsed into his arms.

The newspapers picked up the story and printed articles about the mystery shooting of the artist's model, but did not name Mrs. Epstein as the assailant. Kathleen kept silent and did not bring charges, apparently because Epstein asked her not to. He felt honor-bound to protect his wife and presumably did not want to occasion any scandal which might have been detrimental to his work. A short time afterward, Mrs. Epstein invited Kathleen to drive around Hyde Park with her in an open taxi, so that all the world could see there was no enmity between them. Why Kathleen agreed to this is unknown: perhaps it was to please Epstein. The public charade in the taxi went ahead, but the two women did not meet again.

Prior to this, Epstein visited Kathleen in the hospital and paid her medical bills. Mrs. Garman came up to London to sit beside her daughter and, with her usual unjudging sweetness, didn't ask any questions. Kathleen's shoulder was badly scarred, and she was never able afterward to wear sleeveless dresses, although in later life she would sometimes show the scar to close friends, as if it were an old war wound. In the end, the whole incident became a favorite family story. It was no longer mentioned that Douglas, down from Cambridge, had been staying with Kathleen at Regent Square at the time, and that he had probably chaperoned her to the Epstein house at Guilford Street and waited outside for her. Instead, his convenient presence became the most extraordinary coincidence, that he had just happened to be passing the very door at the very moment when she came staggering out. Another embellishment was that she had been standing over a bowl of deep crimson roses at the moment the pistol went off, and when she looked down, she thought for a moment that the flowers were dropping their petals onto the table, before she realized that the deep red marks were, in fact, her own blood.

That Kathleen later colored the story of the shooting does not diminish the fact that she was, in the event, extremely brave. A young woman of twenty-two, she maintained her dignity and her silence despite considerable pain. Epstein was a controversial figure, and outside the rarefied circle of London's high bohemia, she would have been censured for her association with him. Even within her own family, she was to be subject to opprobrium. Had the truth about the shooting been made known, it would have caused outrage, both publicly and privately, within the Epstein and Garman households. And it is worth remembering that Kathleen's tendency to romanticize was only for the delighted ears of friends and family. To kiss and tell in public would have been completely out of the question for her. Kathleen was a very private person and devotedly high-minded. Her respect for Epstein's work was unwavering, and it always, without exception, took precedence over the claims of her own circumstances.

A few months after the shooting, in the autumn of 1923, Kathleen discovered she was going to have a baby. This, too, would have required a great deal of courage on her part. She would have needed to make herself indifferent to the opinion of the wider world, because the stigma then attached to illegitimacy, both to mother and child, cannot be overstated. While Dr. and Mrs. Garman had been most unusual in their kindness to unmarried mothers in Victorian Walsall, such tolerance was very far from widespread.

During the First World War, attitudes toward "war babies" relaxed slightly for a short time, largely as a result of the efforts of the National Council for Unmarried Mother and Her Child, founded in 1917. One Conservative MP was even prepared to advocate the temporary relaxation of the bastardy laws so as to ease the path of children of men fighting in the war. Many illegitimate children were born during and just after the First World War: between sixty and seventy per thousand births. But tolerance toward them did not last, and the illegitimate birthrate fell back to only four to five percent once the war was over. The belief that unmarried mothers were often mentally subnormal became current; either that or they were morally tainted, corrupt, and therefore a menace to society. The National Vigilance Association was created to put single, pregnant young women in touch with adoption charities, who often levied a punitive charge. In the early 1920s, some orphanages charged fifty pounds to take a child. Such a sum was meant as a deterrent against further promiscuity. Voluntary organizations considered it their duty to inculcate a proper sense of remorse in the mothers. To be a bastard was to be born in shame. It signaled social exclusion, no rights to inheritance nor to the father's name, and disgrace for both mother and child.

These attitudes were not prevalent in the artistic avant-garde, with whom Kathleen and Epstein associated, but among the rest of society such prejudice was to last long into the twentieth century. Just how much people disapproved of illegitimacy may be deduced from the welcome Kathleen received from her aunt Mabel, many years after the births of her children. Kathleen, by then a highly respectable woman with a title, was visiting Herefordshire. Her aunt lived in nearby Weobley, so Kathleen decided to pay her a call. Mabel Hutton was Dr. Walter's favorite sister, and she had also been the close friend and early confidante of Marjorie. In one of Dr. Walter's early love letters, he tells his future wife,

> I gave Mabel your message, and we lovingly talked of you, and appreciated you, together—for she can tell me any little things about you that come into her head—and I can do the

same to her—& we both appreciate them! Today she was talking to me about your schooldays. She says she knows how it is that we get on so nicely with you—and thinks it is because she and I have the same *chemicals* in our nature, and that you like the same in both of us. Anyhow, *you* greatly satisfy us—and we *love* you.

Given such tender mutual regard, it would have been reasonable for Kathleen to have expected a warm greeting from her aunt. She drew up unannounced in a car with a driver and knocked at the door. When a maid answered, she asked the girl to tell Mrs. Hutton who she was. She was left to wait on the doorstep. In due course the maid came back, having been instructed to say that there was no one at home. Because she had had her children outside marriage, and with a married man, Mabel would not allow her own niece over the threshold.*

It was in the summer of the shooting that Douglas Garman left the university. While a student at Cambridge, he had switched from the study of classics to English literature. Books and politics became his greatest interests. Even as a young man he was intellectual and highly serious, although the poetry that he wrote also speaks of a very romantic disposition. He was handsome, especially in profile, when he looked rather like

---

* Kathleen never told this story. Mrs. Hutton relayed it to her children, proud to have seen her niece off.

Douglas in his final year at Cambridge.

Rudolf Valentino. He had a sweet, lopsided smile, and he was dark and tall— six feet two or three—with a characteristic long stride and an easy elegance. He never had any difficulty in striking up a conversation, whether at a party, in a railway carriage, or across a garden fence. His interests were academic, but humor was essential to his nature. Like his siblings he was a brilliant, merciless mimic, and he enjoyed teasing, although he could be prickly.

"He teased everybody, but he didn't like to be teased," his step-daughter remembered. Having grown up in a big family, Douglas was used to ragging, but he could seem harsh to other people. Some found him sardonic, especially in later life when he was sometimes to be engulfed by disappointment, but he was essentially kind. He had integrity.

Douglas wanted to be a writer despite the disapproval of his paternal grandfather, who urged him to join one of the professions. There were only five paths open to a man, old Dr. Garman pronounced: the army, the navy, the law, the church, or medicine. When his grandson said that he was not planning to follow any of them, he was cut off without a penny and the promised support for his younger brother Mavin's university education was withdrawn. Douglas used to sigh that he had been brought up to live as a gentleman but deprived of the means to do so.

If the conservative old man disapproved of the career Douglas intended to follow, his grandson's politics would have horrified him. The Communist Party of Great Britain was formed in August 1920, with ten thousand members. Douglas was interested in the party from the beginning, although he did not join until the 1930s. But unlike many who were attracted to the left then, he remained a card-carrying Communist for the rest of his life, even though his views were not always in accord with the Party's official policies.

At Cambridge, Douglas had made friends with another young man whose political sympathies were in step with his own. This was Ernest Wishart, the son and future heir of Sir Sidney Wishart, Sheriff of the City of London and owner of an estate in Sussex, consisting of ancient woodland and marshy grazing land, close to the sea near Arundel. Always called Wish by his close friends and family, no one who knew him had a bad word to say about Ernest Wishart. He was a gentleman in the true sense of the word: courteous, honorable, modest, and kind. He was extremely generous, and he did not keep a tally of his good deeds. An expert botanist and ornithologist, he was an early conservationist. He was cultivated, eccentric, and very widely read. When he made friends he kept them for life.

Douglas took Wishart to Black Hall to meet his family in the summer of 1925. Lorna was by now an outstandingly beautiful girl of fourteen, with a perfect heart-shaped face, huge Atlantic blue eyes, and endearingly gappy teeth. Ahead of her years and wild, she seduced the much older Wishart in a hayrick ("she was never innocent," as her daughter affectionately remembered). If she was looking for a way out of Herefordshire and poverty, she had found it. Lorna and Wishart were married in 1927, as soon as she was sixteen.

This left only Ruth at home, the dashing Mavin having already run away to sea. The Garmans often quarreled with one another, but everybody loved Ruthie, and she was her mother's favorite. She had been born prematurely and was a delicate child, which carried into her youth and adulthood: she was less good at looking after herself than any of her siblings, more vulnerable, sweeter. Without brothers or sisters at home to check her, she was soon spending more time than ever riding on the back of boys' motorcycles, and dancing and drinking and staying out late. Lorna always felt guilty for having abandoned her former partner in crime, although it was Ruth who was two years the elder. In the early years of her marriage, Lorna would appear at Black Hall laden with expensive presents for her sister.

Douglas, also, had fallen in love, with Mary's friend Jeanne Hewitt. Mrs. Garman was especially fond of Jeanne, and both mothers (each the widow of a Walter) were delighted when the engagement was announced. In London, the Hewitts and the Garmans were constantly in and out of each other's houses and rented rooms. The Hewitts numbered Robert Graves and Allen Lane, the founder of Penguin Books, among their close friends. One sister, the beautiful, doe-eyed Lisa, made her first appearance in the Garmans' lives when she allowed Mavin to kiss her during a visit to Black Hall. He sent wildflowers to her in London and pined, but she paid him no more attention. Later on, Lisa and Jeanne were to reappear in Roy and Mary Campbell's lives in a dramatic way.

As poor as ever, Roy Campbell now decided that he should go to South Africa, where he had relations he thought might come to their aid. Campbell probably liked the idea of returning home in a blaze of literary glory, too. A little bumptiousness can be forgiven: he was still only a young man of twenty-two, and he had not seen his birth family for more than five years. Mary and the infant Tess went to stay at Black Hall for a time, while Roy sailed for Durban in May 1924. "I will hate leaving my two girls," Roy wrote to his mother, just before his departure, "but I think that a trip home will set me up again...I am looking forward to seeing you all again. After all Mary and I have years to be together so we will not grudge the few months that I spend with you." His return was headline news in the *Natal Advertiser*: "A South African Poet: Author Visits His Home At Durban." "It is wonderful, kid, the way they are fussing about me here," he wrote to Mary.

His wife was evidently sad without him. "Don't read gloomy

books," he wrote to her, "just drug yourself with unconscious-
ness as I am doing till I see you again. You must only think of
me in the happy way, as being your lover and Tessie's father.
When I think of you and her I feel that I am a king in spite of
my temporary abdication. I feel no interest in the success of the
Terrapin or anything else, except that I am the lover of Mary
Margaret, the 'Zulu-haired Mary' with a heart like a mountain
boulder and a spirit like a ship in full sail." He had traveled with
a lock of her hair and a photograph, but wrote that she was too
real a presence to him for him to need such souvenirs.

He was soon missing Mary terribly.

I am sick of hanging about without you... All the lovely
things I see are only half as lovely as they would be if you were
here to share them. You have taught me to look at things in the
same way as you do and I *do* miss half their beauty when you're
not there. D'you remember when I used to laugh at you for raving about sunsets etc. I always had the logic on my side but you
had something stronger. What an awful stupid intellectual bore
I must have been in those days. You wonderfully natural creature, it nearly breaks my heart with shame that I ever poo-
pooed your ideas. You are far above me in most things...

Oh my Mary how I bless the day when I first set eyes on you!
I love you more deeply than I ever thought it possible for a human being to love another. Darling how sorry I am now for
getting annoyed with you for feeling sad. I must have been in-
considerate and I'll never be like that again—but all the time I
loved you sweetheart more deeply than I can say: You would

understand how I love you if you could see me now. I am crying like a baby and I must be writing an awfully bad letter. Sometimes I become almost delirious about you and have to ride away alone for miles and miles. I often talk silly love-gibberish to you when I go out into the country alone. I sit on the roof for hours and hours and try to call back every little line of your eyes and face, and think of how we used to lie through the long evenings drowsing together and me kissing you on the back of your darling curly head and the side of your neck, and one arm folded around your beautiful tender breasts: and little Tess snorting in her cot...

With fifty thousand *real* kisses to you and twenty five thousand fatherly ones to Tess,

Your boy,

Roy

Once Campbell was confident he could make a living there, he decided to stay on in South Africa, and Mary and Tess sailed to join him, arriving in December 1924. They had been separated for seven months. Before her arrival, he wrote, "What joy it will be to catch a glimpse of your bonny faces again. How I look forward to our first kiss. I can hardly contain myself when I think of our house in the bush where we will be all alone, away from our relations, away from everything but love and work—the only things that matter at all in life." Before long, he moved his little family out of his parental home in Durban and into Peace Cottage, a bungalow north of the Umthlanga Lagoon, where he had spent holidays as a boy. The walls were soon covered with drawings, as those at Ty Corn had been, leading one visitor to remark that it looked like a Paleolithic cave. There were silvery monkeys in the bush behind the house, and at night the darkness was pricked by fireflies, which Roy sometimes wove into Mary's long hair as they rowed in a little boat together on the warm waters of the lagoon.

But Roy was uncomfortable about being dependent on his cousin, to whom the house belonged, and before long they moved to a seaside bungalow at Sezela. Like Peace Cottage, the house was fairly primitive. It was made of corrugated iron lined with wood, with a shaded verandah on three sides, where they could sit and look out at jungly bush on one side and turquoise waves on the other. They were often to set up home in ramshackle houses like this, between the woods and the water.

The Campbells began to make friends. In June 1925 they met an aspiring young novelist called William Plomer, in Durban. He was so struck by two remarks Campbell made in the early days of the friendship that he wrote them down. "I don't mind repeating other people's secrets, because they are the first to give them away," and, even more tellingly, "One must be theatrical at all costs." Mary certainly shared her husband's relish for drama and enjoyed attracting attention. To go out in the evening she would wear a long red velvet gown, with her hair loose down her back, which made heads turn in a country where respectable women still wore their hair pinned up.

At this time, too, the Campbells became close to a tongue-tied young Afrikaner journalist called Laurens van der Post. "You're one of us," Roy told him. "Come along."

Chapter Four

Magazines

Roy Campbell's invitation to the young Laurens van der Post was not whimsical. He had a specific project in mind; a magazine. It was to be a literary, political, and cultural review, loosely based on the *Dial* or the *Adelphi*. Its name was to be *Voorslag* ("Whiplash"), and he needed the help of his new young friends to get it started. In the spring of 1926, the Campbells tried to persuade William Plomer to be joint editor and to move into a bungalow just by their new house at Umdoni Park. "I am expecting a baby at the end of the month," Mary wrote to Plomer in her characteristically large and open hand, "but if you will risk your nerves with an infant's squeals please come. We shall give you the studio to sleep in and use—I hope you will stay as long as you can. For the sake of the magazine it is important for you to come as Roy is very much in the minority here, really, in spite of all his well-wishers, and he badly needs your backing up. Everyone here is so dead...about anything interesting."

In addition to wanting Plomer to assist with the editorial work, Mary may have hoped that he would help her to manage her husband. Life with Campbell was sometimes difficult. He was extremely sensitive, a hypochondriac, and a drinker. Luckily, Mary had an unwavering belief in his poetic gift. As their daughter Anna put it, "she hoped to find in her young genius an eternal tribute to her own wonderfulness." Mary wrote to Plomer, "Roy wants to write to you, but cannot possibly. He worked all day yesterday from 11 to 8 o'clock with about half an hour's break, with the result that he now has a horrid attack of nerves. These extremes are painful to watch." Plomer later wrote an account of Campbell which fits with Mary's, although he admits a note of skepticism. On many occasions he watched Campbell enter a state of acute neurasthenia, but he had his doubts as to whether these maladies were real or imagined. Some of Campbell's habits were very strange. He ate scantily and irregularly, took no exercise but a great many baths, and was always eating lemons and smoking cigarettes.

Mary became closely involved in the production of *Voorslag*. "Lewis [Lewis Reynolds, their financial backer] blew in! Yesterday morning early," she wrote to Plomer, "he does it so gracefully clad in immaculate white flannels and silkiest of shirts. One expected him to bear a classical scroll in his hand, but instead he had what we almost mistook for a sort of art school circular, in other words the advertisement for *Voorslag*, with a drawing of a buck suitable for the tiles in a ladies waiting room ... Roy says come down quickly, and help with this painful birth of *Voorslag*. We don't want an abortion due to the unskillful interference of other people." In another letter, having congratulated Plomer on the skill of his novel *Turbott Wolfe*, she goes on. "Dear William, as I write I am very upset about Roy and have just woken up from a nightmare of a night, so forgive my ineloquence. I am so glad you are coming and was so glad of your lovely cheering letter. You must help me..." This phrase—"you must help me"—is the only recorded example of Mary displaying the anguish Campbell sometimes brought her. Wyndham Lewis later observed that Mary was imperturbable about her husband's excesses.

In May 1926, the Campbell's second daughter, Anna, was born. She was baptized with water from the Indian Ocean poured over her head from an exotic shell found on the beach. *Voorslag*, too, came into being. The first issue contained Campbell's review of T. S. Eliot's *The Waste Land*, along with extracts from Plomer's new novel. The nineteen-year-old Laurens van der Post also contributed to the magazine, thereby making *Voorslag* a trilingual publication, with writing in Zulu (in which Roy was fluent), English, and Afrikaans. Van der Post used to visit the two founding editors at weekends, taking the train on Saturday night and walking miles to their retreat. The three friends would work and talk, walking along the miles of empty white sand. It was a blissful period that van der Post always remembered with affection.

Campbell could cast a powerful spell. Decades later, at eighty-nine, van der Post recalled a cloudless spring night, cool and clear. To soothe the restless little Anna, Campbell suggested that they walk up and down the beach with her. While they walked, Campbell recited his new long poem, *Tristan da Cunha*. Out on the sand beneath the stars, Campbell at moments seemed to be in danger of dropping Anna, because he wanted so much to follow his habitual gestures, waving his hands when he spoke. Van der Post remembered how the sound of his voice was caught in the roar of the great swell of the ocean breaking on the beach, and foam and spray from the waves came out of the dark with an unworldly glow. To the impressionable younger writer it was as if they walked trampling the stars and the Milky Way under their feet. When at last Campbell came to the end of his long lyrical poem, van der Post was in tears. "It was a moment in my life which has never dimmed," he said.

In England, Douglas Garman also became involved with a magazine. This was to be the distinguished and highly influential *Calendar of Modern Letters*. The poet Edgell Rickword was chief editor, with Douglas as assistant editor, and the magazine was backed by Ernest Wishart, who was then starting his own publishing imprint. Whenever the *Calendar* is discussed, the fact that Wishart was married to Douglas's sister is always mentioned, suggesting a certain amateurishness, as well as hinting at nepotism in his appointment and obligation on the part of Wishart. In fact, Lorna and Wishart did not marry until 1927, and the magazine first appeared in March 1925. It was true that the business manager was Rickword's cousin Cecil, while Mavin Garman was, for a time, the not very efficient advertising manager. But Douglas Garman had the job on his own merits. In addition to becoming an astute and rigorous critic, he was writing poetry. He already cared passionately about the relationship of literature to politics. Over time, he grew to embrace the idea that writers of conscience cannot stand apart from the struggle toward a more just society. As Rickword said in a radio interview in 1967, "We all felt to some degree that literature must be understood and practised as a part of a culture wider and deeper than any single art form, because culture was the essence of the way in which people lived and thought and felt."

Edgell Rickword had been the youngest of the trench poets of the First World War. Serving as an infantry officer, he had lost an eye through wounds. He was slight and fair-haired, with a very quiet manner and a soft voice. His post-war lyrics, the erotic poems in particular, show a debt to Donne and the Metaphysical poets as well as to Baudelaire and the Symbolists. Rickword's political development was typical of many of his generation: converted to socialism in boyhood, partly by his reading of William Morris, George Bernard Shaw, and H. G. Wells, he was confirmed in his beliefs by his wartime experience. He met Douglas Garman in the autumn of 1921, through Roy Campbell, who had been at Oxford with him. Rickword invited Douglas to tea, but they were both too shy to make conversation and ate in embarrassed silence. It was only when they stood on the doorstep, to say good-bye, that they began to talk. It was soon clear that their literary interests coincided. It was to be a lifelong friendship.

By January 1925, they had rented three rooms at the top of

Featherstone Buildings, near Holborn, as the *Calendar's* office. There was a pub across the road, where Rickword would meet contributors and smoke his pipe. Visitors remembered that he spoke so quietly they could hardly discern his words. Nearly half the contributors were under thirty, and Rickword later wrote that the editors considered it their job to "clear the ground, not of weeds only, but some respectable overgrown trees that were depriving the more worthy newcomers of the light which they required to develop." With just such a purpose, a series of "scrutinies," reassessing writers of note, were to become an important part of the magazine. In one such article, regular contributor D. H. Lawrence wrote a damning piece on John Galsworthy, author of *The Forsyte Saga*. The *Spectator* denounced the article as "revolting in taste and indecent in expression." Aldous Huxley, Wyndham Lewis, E. M. Forster, and Bertrand Russell were among the other contributors, as were Rickword's former Oxford contemporaries A. E. Coppard and L. P. Hartley. The composer Cecil Gray, a close friend of Jacob Epstein's, wrote regularly about music. Poetry was supplied by Siegfried Sassoon and Edmund Blunden and Robert Graves. But a piece by James Joyce, the "Anna Livia Plurabelle" section of *Finnegans Wake*, had to be scrapped after the printers refused to set up a passage which they considered obscene. Garman and Rickword themselves wrote for the magazine: in the first four issues at least a quarter of the contents was provided by them and their associate Bertram Higgins, who would ultimately replace Douglas as associate editor. In the early editions, Douglas published an essay on Edgar Allan Poe and gave favorable reviews of Huxley's *Those Barren Leaves* and *The Constant Nymph,* by Margaret Kennedy. He also contributed poems.

A new generation of writers had been coming to prominence, Eliot, D. H. Lawrence, Ezra Pound, James, Joyce, E. M. Forster, Katherine Mansfield, and Wyndham Lewis among them. *The Cherry Orchard* had its first successful staging in England during 1925, and the first translation of Chekhov's earlier piece *The Wood Demon,* was published in the *Calendar.* Virginia Woolf's protofeminist *A Room of One's Own* was to be published in 1929, while Henry Green and Patrick Hamilton were among the more radical new novelists of the time. This was Modernism. It was a time when writing, and writing about writing, had newness and purpose, even urgency. Malcolm Bradbury later wrote that the growth of the modern movement in criticism owed a great deal to the *Calendar.*

F. R. Leavis greatly admired the *Calendar* and used it to teach students. *Scrutiny,* founded in 1932 under the principal editorship of Leavis, was clearly much influenced by it. The magazine also had a strong impact in America, especially on cultural periodicals, such as the *Kenyon Review.* When Malcolm Bradbury wrote his assessment of the *Calendar* in the early 1960s, he compared it to two other reviews of the period: the *Criterion,* edited by T. S. Eliot, and John Middleton Murry's the *Adelphi.* Bradbury noted that it had a coherent sense to it, a conviction about literature, and yet it was open to larger social and moral issues. He concluded that the *Calendar* was much the best, with a fresher, more pragmatic style than the other magazines.

Douglas and his wife, Jeanne, were then living in Bloomsbury, at 10 Millman Street, in a flat which belonged to their friends John and Dorothy Holms. It was here that their only child, Deborah, was born, on January 5, 1926. Douglas's sister Rosalind, always the most practical of the Garmans, was present and helped at the birth. The first visitor was the Ballets Russes principal dancer and choreographer, Diaghilev's successor, Léonide Massine. In the time-honored fashion of those who do not have a crib, the infant was put in a drawer to sleep.

Early that spring, Douglas suggested that he and Rickword could save money on their rent by moving out of London. Rickword's girlfriend, Thomasina, Jeanne, and baby Deborah would go with them. Mavin found two empty cottages at Penybont in Powys, not far from Mrs. Garman at Black Hall. Penybont was very spartan. For a ballet dancer like Jeanne, used to the bright lights, the situation was less than ideal, according to Mavin. Rickword's girlfriend was no more cut out for country life, having been a London prostitute, but her colorful stories kept them all entertained. Hardly anyone thereabouts had a wireless, and there was no news agent in the village. When they wanted word of what was going on in the world, they had to walk five or six miles to Llandrindod Wells. This was how they learned that the General Strike, called on May 4, had begun.

In the years since the end of the First World War, there had been a number of strikes among engineers, railwaymen, cotton workers, miners, foundry workers, and even the police. Such action, however, was not always driven by pay demands: 1920 had seen industrial action from the London dockers, who, in solidarity with their socialist brothers, refused to load munitions onto boats bound for Poland, at that time engaged in fighting against Soviet Russia. In 1923, unemployment reached eleven percent, with many of the unemployed being former soldiers who had to put up with bad housing, inadequate food, and only a pittance of dole money. The General Strike was nominally sparked by the mine owners' insistence on lower wages for longer hours, but this more widespread grievance lay behind it. Garman and Rickword were keen to do their bit to assist the strikers, but it was difficult for them to get to London, as most railwaymen had joined the strike, too. On May 11, they managed to find a train run by scab labor, which got them as far as Crewe, where they were obliged to spend the night in a waiting room full of drunken strikebreakers. By the time they arrived in London the next day, the TUC (Trades Union Congress) had called off the strike.

The workers felt that they had been betrayed by the cowardice of the union bosses. The failure of the General Strike led many who had supported it to commit themselves more deeply to the furthering of the radical cause. It was a formative experience for some writers of the time: Leonard Woolf remembered it as "the most painful, the most horrifying" of British political events to have occurred in his lifetime. Some believe that the cultural dissent of the 1930s actually developed in the midtwenties, with the strike as its catalyst. In any event, when Douglas reviewed Trotsky's pamphlet *Where Is Britain Going?* for the July issue of the *Calendar*, he gave it a sympathetic reading. Since peaceful action had failed, he was led to wonder whether "a regeneration of intelligent sensibility may only be possible after a devastating and bloody revolt against the sickly, bourgeois, animal consciousness of our age."

Douglas Garman now realized that if he was to make a serious study of socialism, he would have to go to Russia to investigate for himself the effect of revolution on culture. Deborah, still a babe in arms, was left at her Hewitt grandmother's in the capable hands of a Welsh girl called Fanny, whom her parents brought to London from Wales. In early November 1926, Jeanne and her husband sailed for Leningrad, where Douglas was able to support them by giving English lessons. They stayed

Douglas (with a Lenin beard) with his wife, Jeanne,
bidding their daughter, Deborah, good-bye, just before they
set off for Soviet Russia in 1926.

for six months. Bertram Higgins was drafted in to fill Douglas's shoes at the *Calendar.*

The magazine overlapped with Jacob Epstein's circle when Edgell Rickword wrote a piece rebuking the "scurrilous vulgarity" of the press campaign against *Rima*, Epstein's Hyde Park memorial to W. H. Hudson, author of *Green Mansions.* Unveiled in 1925, the piece was the subject of widespread loathing. The *Daily Mail* even ran a campaign for it to be removed, under the headline: "The Hyde Park Atrocity—Take It Away." It was repeatedly vandalized, daubed with paint and slogans. Some of this antipathy was aesthetic, but some masked anti-Semitism, as perhaps did Ezra Pound's barb, "Epstein is a great sculptor. I wish he would wash." Cecil Gray, music critic for the *Calendar*, contributed a defense of his friend in which he said that no artist of that or any other time had been so consistently maligned as Epstein.

$\mathcal{B}$y August 1926, Roy Campbell's magazine in South Africa was running into difficulties. Some readers were offended by *Voorslag's* politics, which reflected the editor's delight in courting controversy by openly criticizing white superiority and complacency.* Nor did its generally mocking tone endear it to the readership. The magazine's chief backer, Lewis Reynolds, became alarmed at the way in which *Voorslag* was be-

---

* Campbell's fierce opposition to the color bar is worth stating in view of his later being perceived, wrongly or rightly, as a fascist sympathizer.

coming a mouthpiece for Campbell's and Plomer's revolutionary views, not least because he had more reactionary political ambitions of his own. After some debate, Campbell resigned. He and Mary were left without a regular income, but this was to be one of the most poetically charged periods of his life. The poem *Tristan da Cunha*, greatly admired by T. S. Eliot, came out of these weeks, before the family left their bungalow by the sea, which had come with the job.

Laurens van der Post offered to put them up in his small flat, but it was more practical for them to move to the Campbell family home, now standing empty since Roy's father had recently died and his mother had gone abroad. "We see nobody except a very stupid and nice person called Luther who talks about fishing to Roy and keeps him calm," Mary wrote to Plomer, who had by now become "My dearest William." From here, too, she wrote long letters to her old friend and now sister-in-law, Jeanne. "She is going to Russia next month," Mary told Plomer, "and is going to meet Trotsky [a meeting which never took place] and a lot of other interesting people." Van der Post, meanwhile, was about to sail for Japan and invited both Campbell and Plomer to join him. Plomer went, but Campbell did not want to be apart from Mary. "I should never be half the writer I am, I'm afraid, if it weren't for her," Roy told a friend. "She positively keeps me alive. She is the perfect model of what an artist's wife should be. But I get all the damned credit for it."

When his friends and erstwhile colleagues left South Africa, Roy felt very much alone. The sense of isolation resulted in a breakdown, during which he began to suffer from fits, and he was soon wondering if he was epileptic. He began to write desperate begging letters to friends, hoping to raise the money to pay for the family's passage back to England. "Mary is trying to find work as a lady-help and I have a very bad lesion in my spine which I can only get cured in England. Therefore I am asking you to send me fifty pounds. I shall never be able to repay it. I am asking it as a gift... I am too ill to work. We are going to put the children in a crèche..."

The letters must have had some effect, for in late December they were able to embark for England. The journey was not a success. Mary was seasick and struggling to cope with the two little girls in a tiny third-class cabin. Roy began to drink heavily, depressed by the failure of his return to his homeland. When Mary asked him to look after baby Anna, he would tie her pram to the mast with his belt and go off to drink beer with a pair of cardsharps he had befriended. The passage was so stormy that one stewardess assembled the third-class passengers on deck to say the rosary. Worst of all to the poetry-worshipping Campbells, Roy left the porthole open one day and the long love poem to Mary he had been writing was destroyed by the water which had risen several inches inside their cabin. When the ship docked at Southampton, they were met by Mrs. Garman and Miss Thomas. "Thank God! At last I'm home!" exclaimed Mary.

The Campbells spent a couple of weeks in London before going to stay at Black Hall. Once ensconced in Herefordshire, Mary wrote to Plomer,

I have been in bed for three weeks with a poisoned leg, and am grinding my teeth with rage and impatience at having to be here when there are so many things I wanted to do. We spent a fortnight in London and met almost everyone we ever knew there, but I got ill and had to come back here. Roy was only too glad. He hated London, but although most of the people are fairly hateful, everyone was so frightfully glad to see us that I could not help liking it immensely, and London itself is the most delightful place on earth. I cannot describe the childish feeling of pleasure I always have on going back there . . . all my family have grown up a lot since I last saw them . . . O William the relief it is to be out of South Africa. How I *love* England.

Campbell began to receive twenty pounds a month from his father's estate, and he earned extra money for reviews and poems, some of them published in the *Calendar*. In keeping with its leftist principle, the magazine paid contributors very well, but the expenses outweighed the income. Although it was a critical success, the magazine was not selling enough copies to pay its way, even though it had turned from a monthly into a quarterly. In 1926, new names were added to the distinguished list of contributors: T. F. Powys, Allen Tate, and Hart Crane among them. Rickword had wanted to publish an entire English edition of *White Buildings*, Crane's first poetry collection, but printing problems stopped this. Nevertheless, Crane wrote to Rickword in January 1927, "Probably no-one should be 'thanked' for taking an interest in poetry, but your kindness and interest in what little I've so far accomplished are much appreciated. It is reassuring to me..."*

Despite the high quality of its contributions, the *Calendar* folded in July 1927, partly as a result of financial considerations, partly from the disillusionment of its editor. Edgell Rickword was becoming more politicized and had come to the conclusion that the proper preoccupation of writers must be social change. The questions might be literary, but the answers, he felt, were political. Ernest Wishart continued as a publisher of books, and both Rickword and Garman subsequently did editorial work for him. Wishart & Company was to publish work by Nancy Cunard and Mary Butts, among other left-wing and experimental writers. And it brought out Douglas Garman's only book, a collection of poems, in 1927. The book's title was oddly prescient of Garman's later life, although he could not have guessed it as a young writer in his midtwenties. It was called *The Jaded Hero*.

---

* Hart Crane remembered Rickword's enthusiasm for his work, and they met when Crane visited England in late 1928. Rickword recalled: "It was New Year's day 1929, or thereabout, that I met Hart Crane for the first time. He had been in England a couple of weeks, and had met Paul Robeson [whom Epstein made a head of, and liked] and eaten Christmas pudding with Robert Graves and Laura Riding and was rather miserable... He did not care where we went so long as it did not remind him of the gilt and marble palace in which he dossed, which was as icy as a cellar. But he enjoyed the plebeian sociability of the dockland pubs in Limehouse, though they were not so sinister as he hoped to find them, having seen cinema versions with chinese opium dens and dark alleys where stranglers lurked. And of course we had both been anguished by Lillian Gish in 'Broken Blossoms'...

"Soon we were sitting... facing my recommended ration of two large kippers, two thick slices of buttered toast and a huge mug of tea. The tea was too sweet for our taste, but when stiffly laced with rum from a hip flask, from Crane's hip, it had the aroma of a real Caribbean drink, such as he had been extolling to me, and it blended most happily with the musky kipper flesh."

The Campbells' finances having picked up, they were now able to move away from Mrs. Garman and find a house of their own. One of the editors who printed Roy's work suggested the Kentish village of Sevenoaks Weald, not too far from London. The Campbells moved there, initially to a shack in the woods, close to the poet Edward Thomas's home, until they found a cottage to let. Friends such as the musicians Philip Heseltine and Cecil Gray came to visit, as did Edgell Rickword. Douglas paid an impromptu call, recorded in one of Mary's letters to Plomer: "My brother suddenly arrived on the doorstep last night. He has just come back from Russia and is very nice, much more amusing than he used to be. We were surprised to see him as he and Roy never liked each other and Roy had been slanging his poetry to everyone. We all sat up talking all night. It was awfully interesting."

When Douglas's wife, Jeanne, came to visit her old friend in Kent, she met the Campbell children for the first time. Her first sight of little Anna was of the pretty dark-haired toddler scrambling through a hedge with her knickers down. Five-year-old Tess was meant to be in charge of herself and her baby sister, their mother taking it for granted that they would be able to take care of themselves. Mary's mothering could only politely be described as laissez-faire: to say she was negligent would probably be closer to the truth. She and Campbell instilled a tremendous love of art and nature and poetry in their daughters, but regular meals and bedtimes and the basics of hygiene were not part of their upbringing. They were expected to find their own things to do, without disturbing their mother.

One cousin remembered the young Anna saying, "How lucky you are, that your mother plays with you."

Mary was still corresponding with William Plomer, who was then in Japan with Laurens van der Post. Something of her intolerant nature, as well as her sense of superiority, is revealed in the letters. About Laurens van der Post she says, "I never thought you and Roy would know him long—you wouldn't have done if he had not gone to Japan. It was only the circumstances you met him in that made you notice him at all. You say Van der Post should be a bottle-washer—it is quite unnecessary that he should be anything." (Campbell later fell out with van der Post, telling Plomer that the only reason he'd liked him was because the younger man believed everything he said.) Douglas and Mary were never close, and he does not receive kinder mention. About his newly published volume of poems she reports, "My brother has brought out the most *hopeless* book of verse." Mary always believed that wherever she and Roy were was the best, the most important place to be. "Japan can't be much better than S.A. for anyone staying there—I mean for too long," she writes, later adding, "What an amusing time we shall have when you do come. The English country is too gorgeously lovely. We are in a most English village—the sort I used to imagine I couldn't stand... you can't guess how lovely it is to live here until you have done so."

The tone of Mary's letters changes after this. Writing again at the end of June, she is teasing, arch. "The reason you give me for writing to you is that which most husbands give their wives to induce them to share the connubial couch—my pleasure your duty. I will write to you for any other reason ... What a strenuous regime you have set yourself—do not become monastic I beseech you ... you should give at least part of your time to wine and women—forgive this gross impertinence and label it characteristic instead of something worse ..." Characteristic indeed, for this was written when she was herself giving time over to wine and to women. For early in the summer of 1927, Mary fell in love.

Chapter Five

THE SUMMER SCHOOL OF LOVE

Vita Sackville-West is remembered for two things: the beautiful garden she planted at Sissinghurst Castle in Kent, and her love affair with Virginia Woolf. Photographs show a long woman with a long face. Her habit of wearing feminine, lace-collared blouses with knee breeches led Cyril Connolly to describe her, with neat double entendre, as "Lady Chatterley above the waist and Mellors below." During her lifetime she was well known as a writer, and one book of her poems, *The Land*, won a literary prize, despite being described by Edith Sitwell as "poetry in gumboots." Virginia Woolf was unimpressed by her intelligence but attracted by her aristocratic habits, for Vita was the daughter of the third Lord Sackville.

In a letter to her sister, Vanessa Bell, Virginia Woolf described Vita in an approving rash of semicolons, "very striking; like a willow tree; so dashing, on her long white legs with a crimson bow; but rather awkward, forced indeed to take her stockings down and rub her legs with ointment at dinner, owing to midges—I like this in the aristocracy. I like the legs; I like the bites; I like the complete arrogance and unreality of their minds ... very splendid and voluptuous and absurd ... Also she has a heart of gold, and a mind, which, if slow, works doggedly; and has its moments of lucidity..."

One morning in late May 1927, Mary and Roy Campbell met Vita Sackville-West in the village post office. Vita's curiosity was at once piqued by the attractive couple. The rest of the Garmans had nicknamed them Beauty and Poetry, and the Campbells were still in their twenties, while Vita was six years older than Mary. Roy and Mary were tall and slender and unconventional in their dress, with their own disheveled glamor. Vita had read and admired Roy's poetry, and he knew *The Land*: "It isn't much of a poem, but she is very nice and has a wonderful library...," he told a friend. Vita invited them to dinner on May 27, where they met her husband for the first time. Mary wrote to Plomer, "Our latest acquaintance is the daughter of Lord Sackville. She lived at Knole until she married Harold Nicolson. He has written a quite good book on Tennyson, and she has written a poem called *The Land* which has just won the Hawthornden Prize. They are both rather nice and live in Weald in a beautiful old house crammed with old tapestries, furniture etc. They have some lovely books, and we get quite a lot to read. I have just borrowed Lawrence's *The Rainbow* which is quite the best book he has written... When I was talking to Harold ... suddenly Vita Nicolson appeared, and in her wake, Virginia Woolf, Richard Aldington and Leonard

Mary with baby Anna and toddler Tess, in Kent, 1927.

Woolf. They looked to me rather like intellectual wolves in sheep's clothing. Virginia's hand felt like the claw of a hawk. She has black eyes, light hair and a very pale face. He is weary and slightly distinguished. They are not very human."

During that late spring and summer Roy and Mary and Vita and Harold saw a lot of one another, generally at the Nicolsons' invitation. Many of the important literary figures of the day congregated around the Nicolsons: David Garnett, Raymond Mortimer, and Desmond MacCarthy, then the books editor of the *New Statesman*, among them. Campbell began to feel the guilty rancor of the perpetually obliged, but Mary relished the comfort and grandeur of their hosts' house, Long Barn. Virginia Woolf described a visit there: "Such opulence and freedom, flowers all out, butler, silver, dogs, biscuits, wine, hot water,* log fires, Italian cabinets, Persian rugs, books..." Mary, who had been living in poky and often primitive little houses ever since her marriage, found her head rather turned by all of this. Campbell did not notice extremes of temperature and discomfort, but his wife was more of a sensualist and, like her mother, not fond of housework. Years later, when a relation was telling a story which illustrated someone's cruelty by the fact they had beaten their servants, Mary sighed and replied, "Lucky them to have servants to beat."

On September 2, Mary summoned Vita. "Why did you send for me?" Vita asked her. "Because I wanted you so. There is nothing else to say," Mary replied. Thus their affair began.

---

* An ample supply of hot water was then a great rarity in private houses outside London.

(More prosaically, Mary also needed a lift to the railway station at nearby Sevenoaks. Several of the more dramatic scenes of the affair were to be carried out while fetching and depositing the principals at Sevenoaks station. The Campbells had no car and relied on Vita's motor.) Their proximity allowed them to meet every day, generally in the still-light evenings, when they would walk in the Kentish lanes and woods. They must have looked remarkable: Vita in wide breeches with knee-high buttoned boots worn tight around her calves, Mary in a velvet cloak, with breeches in black or red to offset her dark gypsy looks.

Roy Campbell had once described his wife as a kind of mixture of Sappho and Saint Teresa. Now, with something like religious zeal, Mary called Vita her Saint Anne, her Demeter, lover, mother, "everything in women that I most need and love." Vita liked to take her conquests to see Knole, the vast ancestral home which she had not inherited because she had not been born a boy. Once there, Vita kissed Mary in the bedroom which she still retained, and in her old sitting room they read Shakespeare's sonnets together. "You are sometimes like a mother to me," wrote Mary. "No one can imagine the tenderness of a lover suddenly descending to being maternal. It is a lovely moment when the mother's voice and hands turn into the lover's." Vita and Mary made love on the sofa in Vita's sitting room at Long Barn, the same sofa where Virginia Woolf had once signaled her feelings for Vita. Their passion made them reckless. On one occasion they were within the hearing of Vita's younger son, Nigel, who was sleeping in a room off his mother's, so that she could keep an eye on him while he had the flu.* Mary's little girls stayed with their father during her trysts or were left to fend for themselves.

In September 1927, Vita offered to put the Campbells up rent-free, in a cottage within their grounds. Gardener's Cottage had been used as an annex for the children and their nanny, until the Nicolson boys went back to boarding school. Mary told Plomer, "William you must forgive us for not writing for so long—Roy is busy earning money and I have just acquired a new friend who takes up all my time while Roy is working...we have definitely accepted a cottage from the Nicolsons for about six months I believe..."

At first, the Campbells were glad of their new home, although Roy was later to say,

It was then that we entered the most comically sordid and silly period of our lives. We were very stupid to relinquish our precarious independence in the tiny cottage for the professed hospitality of one of the Stately Homes of England, which proved to be something between a psychiatry clinic and a posh brothel. "Admiration" for my poetry was the pretext for the attempt to exploit our poverty...I was soon in no doubt as to what conditions this "hospitality" entailed. If I have to have vice anywhere around, I like it gay, as at sea, not tearful and tragic; nor do I like it as a sort of obligation for board and lodgings... The one excuse for being a pagan is to enjoy it thoroughly. But these people resembled the old Greeks as a Chelsea Eurhyth-

* The best-selling *Portrait of a Marriage*, Nigel Nicolson's account of his parents' life together, gives only one sentence to the Campbells.

mic Class resembles the Pyrrhic Dance. The Sevenoakians were solemn hyprocrites.

This account was written with bitter hindsight. Campbell was working for various magazines at the time, in order to support his family. "Roy earned £9-9 last month by writing one article on Blake for the *New Statesman* and two poems," Mary told Plomer. "We met Desmond MacCarthy and he is very nice and extremely generous." But Campbell was also drinking heavily. Much as his wife loved England and their new circle, he missed the climate of South Africa and chafed under the gratitude he owed the Nicolsons, both for the rent-free house and for the patrons, such as MacCarthy, he had met through them. At dinners with his neighbors, he became increasingly monosyllabic and morose. He was not writing much poetry, and poetry meant everything to him. Mary would return from the village shop to find the children alone in an empty house, Roy having gone to London to drink with one of his cronies, Philip Heseltine or Augustus John. It is likely that it was partly Roy's drinking which drove her to seek refuge with the older and more settled Vita. And with Campbell in London, she was able to carry on the affair without her husband's knowledge.

The intensity of Mary's feelings for Vita may have been reciprocated, but for Vita the attachment was not exclusive. She was still carrying on an ardent correspondence with Virginia Woolf; toying with an old girlfriend, Dorothy Wellesley (later the Duchess of Wellington); and flirting with a young actress, Valerie Taylor, who had become infatuated with her. Taylor told Vita all her love troubles and remained with her, after dinners at Long Barn, into the early hours of the morning. She clearly hoped to seduce Vita. On one occasion, she went to Long Barn dressed up as Lord Byron.

At the end of October, Harold Nicolson, who worked for the Foreign Office, was posted to Berlin. His wife told him she was quite wretched, she missed him so. Nevertheless, Vita noted in her diary in early November, "A very lovely sunny day; a brown and blue and purple day. Went for a walk with Mary in the morning . . . Roy telephoned to say he would remain in London tonight. I wrote . . . while she played the gramophone in the evening. She dined with me. A very happy day."

That happiness was about to come to an end. The next day, after the two lovers had met Campbell from the London train in Vita's car, Vita went home and Mary told her husband about the affair. Such confessions generally presage the end of an illicit romance, but Mary had no intention of giving Vita up. Perhaps she hoped to imitate Vita herself, who confided in her own husband about all her liaisons. But Roy Campbell was more passionately attached to his wife than Harold Nicolson was to his. At first, he took the news calmly. When the Campbells met Vita out shooting pheasant in the dusk, all three went back to Vita's sitting room, where Campbell announced that he was himself sleeping with another woman. But such bravado did not last the night. Vita's diary for November 7 reads, "A dreadful day. Roy had kept Mary up practically all night with threats of murder, suicide etc. Went on doing my Oxford paper miserably with these constant incursions to tell me of Roy's changes of mind etc. . . . In a rainy dusk I walked up to the village with M. to get beer for him. After dinner unable to bear things any longer I went up to the cottage and talked to him. She, poor child, quite numbed and dead tired."

Campbell retreated back to London on the train, intending to drown his sorrows. Once installed in a pub, he chanced upon C. S. Lewis (who had been a friend at Oxford) and told him the whole story. "Fancy being cuckolded by a woman!" said Lewis. Roy was a proud man, and this remark so punctured that pride that he returned to Kent in a towering rage. A terrified Mary took refuge at Long Barn, where Dorothy Wellesley sat up all night with a shotgun across her knees. Later, Roy commemorated her in a poem, "Gin-soaked, a shot-gun in her clutches / The Fury was a future duchess . . ."

Vita told Harold that Roy had gone for Mary with a knife, but she did not allow the drama to interrupt her social schedule. Clive Bell and his brother, Valerie Taylor, and Virginia Woolf came to visit and were taken over to Knole for the afternoon. That same morning, "Roy came down just before lunch to say they were going to be divorced," Vita recorded, "I told him not to be so silly." Once her guests had left, Vita went to the cottage. Campbell had calmed down, and they were able to talk to each other, which made Vita, at least, feel much better. Roy wrote to Vita, "I am tired of trying to hate you and I realise that there is no way in which I could harm you (as I would have liked to) without equally harming us all. I do not dislike any of your personal characteristics and I liked you very much before I knew anything. All this acrimony on my part is due rather to our respective positions in this tangle. I am much more angry with M."

All parties did a lot of writing in the weeks that followed. Mary tore pages out of notebooks and account books and scrawled frantic messages to Vita on them: "Is the night never coming again when I can spend hours in your arms, when I can realise your big sort of protectiveness all round me, and be quite naked except for a covering of your rose leaf kisses?" Vita had said that the price of their love affair was too high, pre-sumably in the woe it was causing Roy, which elicited further thoughts from Mary. "Darling is it very often that we get weeks of such amazing happiness...and I *don't regret it at any time,* even when I suffer most it becomes more worthwhile." Vita, meanwhile, was inspired to verse by her sense of poetic com-radeship with Roy:

We both have known your beauty, both enjoyed
Your passion, as a shared and secret thing;
On that account, each other we avoid
Who both have tapped your passion at its spring.
Is this a reason? Should we not more truly
Meet in our common love that is your due?
Across your body not discover newly
A bond the deeper for its source in you?

She also wrote a short story, depicting a thinly disguised Campbell as the one truly noble person among them. Despite such lofty renderings of the situation, Vita confided to Harold that it was really "a mess, and nobody is pleased." That December she composed sixteen sonnets about the affair, a sequence that served as a kind of catharsis. After love's first careless rapture, she was coming back down to earth, writing calmer letters to her husband, planting a wood, and having central heating installed. (Of this venture she noted: "And my God how workmen smell. The whole house stinks of them. How I hate the proletariat.") Vita went to Berlin on December 14, where letters from Mary, who had borrowed her car, followed her. Calling her "my rose, my darling mother," she lamented, "you should have seen me sitting in your motor in Piccadilly Circus crying quite helplessly."

The affair continued into 1928, with Vita writing four more sonnets to Mary in February, while Campbell went into the hospital to have his appendix out. Laurens van der Post, who visited him soon afterward, found him in a pathetic state: agitated, shabby, and alone in a freezing house. (During the night, Roy stole in and with characteristic generosity put his own blanket on top of his guest.) His wife's absence was not explained and he was still drinking heavily. Van der Post noticed that when Mary came back Roy cheered up and became much more like his old self.

While Vita visited Harold in Berlin—"a bloody place, to be sure," Vita thought it—she and Mary wrote to each other every day. Despite the exchange of letters, Vita was becoming involved in yet another entanglement, this time with Margaret Voig, an American who wrote and acted as an agent for English-speaking authors. But she soon came back to England, and to

Mary. The two spent Vita's first night together, having met up in the foyer of a London cinema. "My own I still feel warm from your arms," Mary wrote her two or three days later. When Margaret Voig turned up in London a month later, she hoped to stay at Long Barn, but Vita arranged instead for her to be put up with Vanessa Bell in Gordon Square. Perhaps it was Mary's proximity in Kent which made Vita keep her new lover in London. Poor Margaret had been in England for only a week before Vita began to tire of her, telling Harold, "I forget about people very quickly when I am away from them."

In April, the disconsolate Campbell went back to Martigues, a fishing village in the south of France where he and Mary had spent a happy holiday the summer before. He wrote plaintive letters to his estranged wife, begging her to join him: he could not live without her, he pleaded. Although Mary was still deeply in love with Vita, she went to him. Vita took her to Sevenoaks station, waving her off minutes before meeting Margaret Voig (now back in favor) from another train and taking her back to Long Barn for the weekend. According to the Campbells' younger daughter, Anna, Vita now asked Mary whether she could adopt the little girl, "as I was a beautiful child." Perhaps she would have liked the daughter as a souvenir, but Mary refused. The Campbell children were left with a nurse in Kent, while Mary went alone to France to talk about reconciling her marriage. From France she continued to write secret letters to Vita almost daily, apparently going to great lengths to ensure Roy knew nothing of the correspondence. She went back to England to see her lover twice that year: for ten days in June, when she collected the children, and again in November. But Vita's passion for her cooled with the seasons. Although Mary thought again of leaving Roy, she realized that by now Vita would not encourage her to do so.

The affair had a number of literary consequences, the most important being Virginia Woolf's extraordinary novel *Orlando*. It was during the summer when the romance between Mary and Vita was at its height that Woolf, driven by jealousy, devised the outline of the book.* "I rang you up just now," she wrote to Vita, "to find you were gone nutting in the woods with Mary Campbell...but not me—damn you..." She wrote to Vita again in October, "suppose *Orlando* turns out to be about Vita; and it's all about you and the lusts of your flesh and the lure of your mind (heart you have none, who go gallivanting down the lanes with Campbell)—suppose there's the kind of shimmer of reality which sometimes attaches to my people...Shall you mind?" Far from minding, Vita was delighted: "My God, Virginia, if ever I was thrilled and terrified it is at the prospect of being projected into the shape of Or-

---

* A possible influence on the plot of *Orlando* is the novel *The Sundered Streams*, by Reginald Farrer. Farrer is today better remembered as a plant collector than a writer, but according to Nicola Shulman's biography of him, *A Rage for Rock Gardening*, Virginia Woolf gave a favorable review to his first novel (in the *Times Literary Supplement*, 1906). If Woolf read his next novel, its influence on *Orlando* has never been examined. Yet the two bear remarkable similarities. Farrer's novel features a character who starts out as a woman, albeit a mannish and unreserved one. She assures the hero that they have been lovers through all eternity, their souls destined to search for each other through many lifetimes. She is later reincarnated in the body of a young man. Although his influence on Woolf has gone unnoticed, Reginald Farrer's style was acknowledged as an influence on the garden writing of Vita Sackville-West.

lando. What fun for you; what fun for me. You see, any vengeance that you want to take will be ready in your hand... You have my full permission... And what a lovely letter you wrote me, Campbell or no Campbell. (How flattered she'd be if she knew. But she doesn't, and shan't.)" *Orlando* was published in October 1928, with three pictures of Vita among its eight photographic illustrations.

Mary wrote to Vita from France, having read the book, "I hate the idea that you who are so hidden and secret and proud even with people you know best, should be suddenly presented so nakedly for anyone to read about... Vita darling you have been so much Orlando to me that how can I help absolutely understanding and *loving* the book... Through all the slight mockery which is always in the tone of Virginia's voice, and the analysis etc, *Orlando* is written by someone who loves you so obviously... Don't you remember when we imagined you as the young Orlando?" But Mary felt that Virginia Woolf had left one important aspect out of her character: "It is just hinted at by the word 'luxurious'... Orlando is too safe too sexless and too easy-going to be really like you. But I am thinking of him as he appears to *me*, he is something so different to Virginia. Ah! an entire book about Orlando with no mention of her deep fiery sensuality—that strange mixture of fire and gloom and heat and cold—seems to *me* slightly pale."*

---

* Vita's sexual appetite was allegedly prodigious. An apocryphal story had it that Mary came home one day with terrible cuts and bruises on her thighs, at the sight of which Roy is said to have exclaimed, "Good Heavens, kid! I don't mind you sleeping with Vita, but at least get her to take her earrings off!"

Virginia Woolf made no reference to Mary Campbell in her novel, but Vita did in her writing of the period. A collection of poems, *King's Daughter* (the word king punning with Roy) was published in the autumn of 1929. Vita included only three of the many sonnets she had written about Mary, but the erotic content of the work was unmissable. "It has occurred to me that people will think them Lesbian," Vita wrote to Harold. "I should not like this, either for my own sake or yours..." Such rectitude rings false. Vita had had no such scruples about being identified as the heroine of *Orlando*, but since she had not written that novel herself, she did not have to first clear it with her husband. One of the poems included in *King's Daughter* dealt summarily with her affair with Mary:

The catkin from the hazel swung
When you and I and March were young

The flute-notes dripped from liquid May
Through silver night and golden day.

The harvest moon rose round and red
When habit came and wonder fled.

October rusted into gold
When you and I and love grew old.

Snow lay on the hedgerows of December
Then, when we could no more remember.

But the green flush was on the larch
When other loves were found in March.

Roy Campbell, too, was inspired to write by what had happened. His long satirical poem *The Georgiad* did not appear until 1931, although he probably started writing it while still living in Kent. The poem was to cause a furor in the literary world, as Campbell lambasted Bloomsbury, its values, and its characters, the Nicolsons chief among them. Many people are named directly: Robert Graves and Laura Riding, Raymond Mortimer, Desmond MacCarthy, Bertrand Russell. Satire tends not to age well, but parts of *The Georgiad* stand up to time, although it is more often remembered as a literary footnote than read. Mocking the double-gendered Orlando, Campbell creates a "Hermaphrodite-of-letters," whom he calls Androgyno:

My manhood, with unfeeble sensations,
Is changing into ladies' combinations,
This hairy thigh, which pants enclosed before,
Now shivers in a flimsy silken drawer,
Half of your corset round my ribs is locking,
Along my shins there crawls a long blue stocking . . .

Nor is there any mistaking the location he writes of:

The Stately Homes of England ope their doors
To piping nancy-boys and crashing Bores
Where for week-ends the scavengers of letters

Convene to chew the fat about their betters . . .
O Dinners! take my curse upon you all,
But literary dinners most of all,
Where I have suffered, choked with evening dress
And ogled by some frowsy poetess . . .

Long Barn is rechristened the "Summer School of Love," with Vita (the frowsy poetess) as the thickly mustachioed "Georgiana":

Too gaunt and bony to attract a man
But proud in love to scavenge what she can,
Among her peers will set some cult in fashion
Where pedantry may masquerade as passion,

Even the Nicolsons' dog is named and vilified,* along with Vita's gardening. Nor is her poem *The Land* immune from mockery:

. . . safe-sequestered in some rustic glen,
Write with your spade, and garden with your pen,
Shovel your couplets to their long repose
And type your turnips down the field in rows . . .

. . . while in rhyme you keep the farmer's books,
Your soulful face will scare away the rooks,
While wondering yokels all around you sit,
Relieved of every labour by your wit.

---

* Roy Campbell loved animals, but being bitten as a child created a lifelong aversion to dogs.

Campbell was delighted by the sensation which followed publication. As he wrote to his friend Wyndham Lewis (whose novel *The Apes of God* had much influenced him), "MacCarthy Nicolson & Co are paralysed... Since the Georgiad (I hear) the Nicolson menage has become very Strindbergian. Each accusing the other for it and smashing the furniture about: but they are rotten to the core and I don't care about any personal harm I have done them—I take their internal disturbances as a justification of The Georgiad." The Nicolsons, meanwhile, behaved with admirable reserve. "I know he is a very queer character," Vita wrote of Campbell, "but I don't really bear him any grudge, and I shall continue to think of him as a very fine poet. I detest literary rows and will never be drawn into them." But the gossip and rumors muttered on, with many in the world of letters believing that Campbell had simply bitten the hand that fed him. Years later, in response to a professor who was planning a biography of Roy Campbell, Harold Nicolson attempted to have the last word:

I believe it is fair to Roy Campbell's memory to enable his biographer to realize that he was not so mean or vindictive as some people suppose.

The current legend is that out of charity we lent him the gardener's cottage at our home at Long Barn, Sevenoaks Weald. That we there introduced him to several of our literary friends who came down to dine and sleep or to stay from Saturday to Monday.* That some of these friends, notably Raymond

---

* Nicolson is deliberately avoiding the phrase *weekend*, which, like the word *holiday*, was considered vulgar.

Mortimer and Edward Sackville-West (my wife's cousin), did not pay sufficient attention to the Campbells and in fact talked about people they did not know or books in French and German which they had not read. That Roy Campbell was incensed by this behaviour and acquired angered feelings of inferiority. That he therefore quitted the house and thereafter revenged himself on all of us in the Georgiad.

This legend is only partly true and suggests that the invective of the *Georgiad* was attributable solely to motives of trivial spite. But his rage can be explained, and in some sense justified, by another reason.

Mary Campbell at that date was passing through a difficult period and confided to my wife her resentment of Roy Campbell's ruthless selfishness. My wife was moved by her confessions and Mary Campbell (who I can well suppose had not made confidences or received sympathy before) responded with deep devotion. I do not know how far the matter went, or who of the two was most to blame. But at one stage Mary Campbell must have told her husband that her affections were becoming involved and with characteristic violence he dragged her away from the place and broke off relations. Thereafter he must have brooded over the occurrence and fused his jealousy with resentment over what he took to be the social and cultural conceit of the Bloomsbury friends he met at Long Barn. I have never seen Roy Campbell again and in so far as I know my wife never again saw Mary Campbell, although they sometimes wrote to each other.

These are the facts so far as I know them. I have never discussed the episode with my wife, since I know she feels ashamed at having exposed my friends and myself to an invective which was damaging and undeserved. But she has so generous and noble a nature that she bears no resentment against Roy Campbell, whom she regards as a fine poet who was slightly insane and utterly ruthless in taking revenge.

Harold Nicolson sent a copy of this letter to William Plomer: "I think one owes it to the memory of a poet to disclose that the motives that inspired his satire were not solely due to causeless spite. But I don't quite enjoy doing this, since as you know my loyalty to Vita is the pole-star of my life. The whole business is a painful memory to her and I shall not show her the letter. But I think it is right to send it."

Before publishing *The Georgiad*, Campbell had brought out another book of poems, *Adamastor*, in 1930. For this volume of sun-drenched and moonlit lyric poetry, he wrote a dedication to Mary:

> *. . . When the Muse or some as lovely sprite,*
> *Friend, lover, wife, in such a form as thine,*
> *Thrilling a mortal frame with half her light*
> *And choosing for her guise such eyes and hair*
> *As scarcely veil the subterfuge divine,*
> *Descends with him his lonely fight to share—*
>
> *He knows his gods have watched him from afar,*
> *And he may take her beauty for a sign*
> *That victory attends him as a star,*

Shaped like a Valkyrie for his delight
In lovely changes through the day to shine
And be the glory of the long blue night.

When my spent heart had drummed its own retreat,
You rallied the red squadron of my dreams,
Turning the crimson rout of my defeat
Into a white assault of seraphim
Invincibly arrayed with flashing beams
Against a night of spectres foul and grim.

Sweet sister; through all earthly treasons true,
My life has been the enemy of slumber:
Bleak are the waves that lash it, but for you
And your clear faith, I am a locked lagoon
That circles with its jagged reef of thunder
The calm blue mirror of the stars and moon.

Their marriage had survived the earthly treasons of Mary's affair with Vita. Now, in a small stone house on a hill of scented pines near Martigues in the south of France, the Campbells began a new life in the sun.

Chapter Six

WING MY HEART

"What a pity every man of genius cannot have such a wife!" one of Jacob Epstein's models said of Margaret Epstein. In 1922, the ever-resourceful Mrs. Epstein found a house at the edge of Epping Forest, where she and her husband and little Peggy Jean spent a good deal of time, Epstein working in a small summerhouse at the end of the garden. If she hoped that her husband would forget his lover while he worked in the dappled light of the woods, she had made the move in vain. Instead, he wrote to Kathleen,

It is now a week since I've seen you and I've so longed to see you that my whole day and night is conditioned by my great longing. You do well to write in your letter of kisses and embraces that is what I think of and were we together it would only be to kiss and embrace you; you are right also when you speak of our mysterious happiness. I am not altogether wed; I do not go out enough and last night I believe I spent almost the whole night thinking of you... I am not with you my arms that wish to hold you haven't you my hands that want to feel you miss you: yours are the kisses I want...

Throughout the 1920s he still came to stay with Kathleen two or three times a week, always on Wednesday nights and often on Saturdays, and when he was not at Epping, he liked to drop in on her every day for an hour or so after he had finished work. "How I wish we were together you and I in our little place. How happy I would be beyond words and all expression in words...," he wrote from Epping. Epstein was given to appearing with great armfuls of red roses and a bottle of wine, when what Kathleen really needed was money for provisions, including coal for the fire that was the only source of heat in her studio room. He did give her money after selling a piece of sculpture, but it was a very erratic income. At least she now had the company of her sister Helen, who had come to London to join her. Helen also acted as a go-between for Epstein's and Kathleen's letters, and it was she who told him, in a secret rendezvous in the forest at Epping in July 1924, that Kathleen had given birth to a son, Theodore.

Kathleen was mostly alone with their son, while Epstein now had two children in his principal household at Guilford Street. In addition to his daughter, Peggy Jean, there was now a little boy called Enver, the child of Sunita, Epstein's new model. Sunita was an Indian girl from Kashmir and, like Kathleen, a doctor's daughter. She and her sister had run away from unhappy marriages and had settled in England, where Sunita had made the acquaintance of Epstein's friend Matthew Smith and become a model for him

Rosalind, Kathleen with baby Theo, and Helen, spring 1925.

and, possibly, his lover. Before sitting for Smith, she had worked for the magician Jasper Maskelyne,* performing a mermaid act in a glass tank full of water. Almost as soon as he met Sunita, Epstein invited her to move in, along with her son and her sister. He said she was the most beautiful woman he had ever seen. Mrs. Epstein seems not to have minded this incursion, hoping that Sunita would distract her husband from Kathleen. Epstein did become obsessed, but not romantically. Instead, he drew and sculpted from the new model and, to a lesser extent, her son and sister.

While Sunita was his model (for nearly ten years between

---

* Jasper Maskelyne's magic show is still remembered with affection by Patrick Leigh Fermor, who saw it as a child. During the Second World War, Leigh Fermor met Maskelyne in Cairo, where the former showman turned his talent for disguises to the use of the Allies' desert campaign, where he "vanished" the Suez Canal, using reflectors to create the illusion that the canal was not there, thus evading enemy attack. Maskelyne died in obscurity in 1972.

1922 to 1931) he did not make a single sculpture of Kathleen. Kathleen's three children were all born during these years, and perhaps he did not work from her in order to keep any visible signs of their arrival from his wife, who was eagle-eyed about the details of his work. Perhaps, in view of the frank sensuality of the second 1922 head, his wife had abjured him to stop modeling from Kathleen. But he resumed working from her after Sunita returned to India, where she died young.

In 1927, Epstein traveled to New York for an exhibition of his work, taking his wife and daughter with him. Kathleen remained in England. By now she had two children: Theo was three, and a daughter, Kitty, still a baby. From early infancy, Kitty was brought up by Kathleen's mother and Miss Thomas. Mrs. Garman still had her two youngest daughters at home when her granddaughter arrived, and Kitty grew up thinking of her aunts Lorna and Ruth as mentors and honorary older sisters. Coping with a small child in the one-room studio Kathleen was now sharing with her sister Helen must have been difficult enough, and she evidently felt she could not manage a newborn baby as well. Their quarters were cramped and she was very short of money, even though Helen had learned to type and was able to find work to help support her sister and pay the rent on their lodgings. Kathleen did what she thought was best for her daughter by sending her to stay with her mother at Black Hall in Herefordshire. The clean air of the countryside seemed more beneficial to children than London's thick fogs, and theories about the importance of bonding and attachment between mother and child had yet to evolve.

Kathleen did keep Theo with her in London. He went to school in Chelsea, winning a scholarship to Sloane School in

1937. But when her second daughter, Esther, was born in 1929, she was also sent to the country, to Kathleen's former nanny, Ada Newbould. Ada lived in the Black Country at Shelfield, where she and her blind husband ran a shop. Esther lodged there until she was seven, developing a strong West Midlands accent. During the school holidays, Theo would join her. In time, Helen's daughter, Kathy, would also stay sometimes with Ada. "We all loved Ada's," she recalls. "We used to go and have holidays . . . she had a sweetshop at the front: it was so exciting. We'd have lovely things for high tea: Spam and chips and sausages, and fish and chips." To Kitty, who sometimes joined her brother and sister there, Shelfield was a place of dread: deeply attached to her grandmother, she felt no bond with Ada and was frightened of her husband. When the greater Birmingham area was threatened with bombing raids during World War Two, Esther was sent to join Kitty in the comparative safety of their grandmother's house. Not until they were in their teens did either of the girls spend any length of time with their mother. Kitty remembered staying with her mother in London once and being terrified when, the age of seven, she was left alone at night with Theo, while Kathleen went out with Epstein.

The prevailing attitude toward parenthood reveals itself in the marked difference between Epstein's letters to Kathleen and those which Walter had written to his beloved Marjorie when they were apart, some thirty years before. The avant-garde set of the 1920s were not sentimental about children as their Victorian and Edwardian predecessors had been. While the doctor inquired tenderly about little Mary and baby Sylvia, asking that his wife kiss the girls on his behalf, Epstein barely mentions his offspring. Neither, it seems, did Kathleen give an account of them in her letters. "You don't write to me of some things," Epstein noted. "I expected you to write about your visit to Blackhall and perhaps it's in more than bad taste to mention it to you as you so obviously omit it on purpose. You know what I mean when I ask you how our little girl is." This reference to their child is exceptional. Epstein was unusual for a man of his time (as was Roy Campbell) in that he did love babies and children and was natural and affectionate in their company, but the letters show how much he still looked upon Kathleen as a lover and not as a mother. She remained his prize, his secret. Kathleen would have sacrificed anything for Epstein, even the care of her own daughters.

Epstein was absorbed in his own work above all, but the love and desire he felt for Kathleen burned with something of the same intensity he brought to his sculpture. "Apart from working being with you sweet girl has been and is the obsession of my life," he wrote from New York. "Surely anything one does could be good if one worked with sufficient concentration and intelligence. Love I was going to say but it is dangerous to talk personally about art I mean to relate work to oneself and yet that is what I am always doing..." His letters from America also provide clues that he was influenced by Kathleen's eye, as he longs to know what she thinks of their friend Matthew Smith's latest exhibition or of Giorgione or of the Rembrandt cards he sent her from the Metropolitan Museum of Art. "You know sweetheart as well as I do that my things in important instances are suggested by you... The figure at the Tate was entirely suggested by you... I've never confessed to it."

Epstein was unhappy in America, despite receiving enormous press attention.

I could get heaps of portraits and I'd be condemned for the rest of my life to do the faces of American professors and business-men. Curiously enough no women want these busts only men and what I think I dislike most is the type of academic professor: they swarm in thousands and my entire time could be taken up by these dull fellows if I let them... How I long to be back with you again Kitty* with you in all loving ways. How I enjoy watching you holding you and the peace you write of will only be mine after I've seen you again. I am getting more and more irritable and find those I meet tiresome. I think you are never out of my mind... My nights are my worst time. I lie awake and think of you. Last night, Sunday, was wonderful moonlight and I thought of you, picturing how you were, how fascinating you looked on the balcony beside the lake in the moonlight and lantern lights. Very romantic!... I'm glad you enjoyed the orchids and I look at the ending to your letters. I am always yours Kitty.

Although he may have been absentminded about Kath-leen's poverty when they were together in London, Epstein sent money from America whenever he could. On October 14, 1927, he sent fifty-five pounds† or thereabouts, then ten days later another eight, wishing that it were more. "I will sell and make money and at last for my sweetheart I would be able to get those beautiful clothes and shoes I have always wanted to get her..." He was buying her a piano on an installment plan, toward which he sent a further six pounds. She sent him presents, too: a bunch of red roses for the opening of his New

* Epstein called Kathleen Kitty, the same name as their daughter.
† One pound in 1927 is worth roughly forty pounds now.

York exhibition, an antique blue ring for his birthday. "If I send you anything please spend it on yourself and not on me," he answered, worried that she was running short.

As the weeks pass, he becomes more lovelorn. "I long for you as much as you do for me and I brood and think over the past and every detail is beautiful," he wrote in late October.

> I forget and see how small our bad times were: they are nothing and only our happiness remains in my mind and fills me with joy and unhappiness that I am not living again with you. This is a barren time for me. Be patient a little sweetheart, so that I can get this exhibition through...Now that we are so forcibly separated, I feel more than ever that my life is inevitably with you. I have had such bad hours that could I manage it I would have taken a boat and come straight to 13 Regent Sq. and lived with you from then on...I sometimes imagine I'm in some sort of a nightmare and nothing is genuine here. I think constantly of you darling and your letters will wing my heart as I read it [*sic*] over by myself on lonely walks in the park or in a coffee house, where I am writing now. I row on the lake in the morning early, when no one is on it. This in the midst of the city in Central Park amongst the tall buildings. There I can take out your letters and read them. I look forward to the future; it holds my happiness again with you.

A few days later he was writing again, this time piqued by the fact that Kathleen had been out in London.

I have your 3rd letter and how happy I am to send you loving words and your curious little note about the gargoyle[*]
makes me wonder. Now you had to wait until I left London
before you could be taken to such interesting places and I
wonder who took you back to your place despite the fact that
you didn't dance. I never danced either with you at the club but
as you say we were mostly gay and even if not gay always later
madly happy . . . My thoughts constantly, daily and hourly go to
you and I want to picture you going about in London. I'd like
to send you more money than I do and if I make a sale and
money I will. Until then sweetheart I am forced to send you the
small sums I do . . . There is much that is ugly here and it isn't
my idea to stay here . . . I care nothing for these tall buildings
and self inflated ideas that prevail everywhere. There is a remarkable phenomenon here that almost every other man imag-
ines he has "genius." I listen to terrible commonplace platitudes
given out solemnly and I feel like seeking solitude. Oh to be
with you in our small room together happy always happy.

By now, Mrs. Epstein had become terrified that her husband
might leave her. "I was the only person she could confide in," Peggy
Jean later told Epstein's biographer, Stephen Gardiner. Her adoptive mother told her they had to do all they could to restrain him
and make him stay in New York. According to Epstein's sister,
Sylvia, who worked for Wonder Dress, a New York garment firm,
Margaret Epstein was always moaning that her husband was about

[*] The Gargoyle Club in Soho opened in the mid-1920s and quickly became a favorite among London's bohemians.

to leave her, at which Peggy Jean would reply, "Oh Mummy, if he hasn't left you by now, he'll never leave you." Sylvia felt that Margaret was everything her brother needed. "She looked a little like my mother; she was heavy and she was sweet like my mother. In fact, when my father visited them in England, he came back and told me: "She looks like your mother." When Mrs. Epstein lost the battle and Epstein insisted on returning to England ahead of her, she led Sylvia to believe that he had been forced to go back because Kathleen had been threatening to kill herself. In fact, Epstein's unhappiness in America owed a great deal to his father's having destroyed all the early drawings he had left in his (now dead) mother's keeping. He also felt uncomfortable promoting himself as the American art market demanded. In later years, when he visited the grown-up Peggy Jean in the United States, he became much fonder of his birthplace.

Evidently, Kathleen missed him, but there is no hint that she was ever suicidal: "I have been reading your letter your unhappy letter again," Epstein wrote her,

and here in a quiet place in the park I can write you how much I love you and think of you. We are separate now but will be together again joined together and all my images are of you and of our good times together. I hope you are not in want I mean that you are able to pay for your piano and get along and even buy some flowers for your vase although I can only now imagine them. I imagine your room with me in it or is there another to take my place. Do you meet interesting people, new people or old friends and now that I don't take your time have you more time for them? I am really unhappy and miss you terribly…

Kathleen carried on with life in England, dining with Jacob Kramer, meeting visiting Russians, and going to the Café Royal and other haunts. There were weekend visits to her mother in Herefordshire, to see little Kitty. Epstein's New York show had opened, but he was still out of sorts with his surroundings: "They are terrible sentimentalists here but I suspect hard cash and business behind them...My family were terrible and ready to tear each other to pieces over me..." His next letter was no more accommodating: "Americans are mad. Their hot stuffy rooms incessant chatter and bawling appal me and I want to get away." At the end of November he responded to a letter from Kathleen. "My show is nearing its end here. I read of the foggy day in London* in the papers here and thought exactly as you did how close we could get together and how happy we could be on such a day as that when we could shut out the world. I know I am missing all that and I would not try to forget what is unforgettable and that is what I hope you also remember. I'll not write again of future hopes and desires as that irritates you and you think I don't live practically or even think practically and certainly until I am really with you again I can only send out to you what perhaps seems vain complaints and longings..." He goes on to note that "millionaires here only give money to charities and buildings where their name is blazoned...Wait for me sweetheart and don't forget that I am your lover though far from you."

A show in Chicago followed. Epstein made a portrait bust of Paul Robeson, whose head he much admired, beginning a friendship that was to last for many years. His letters to Kath-

---

* London fogs—called "pea soupers" in honor of their thickness—continued until the Clean Air Act of 1956.

leen become more and more conciliatory. "My girl my sweetheart is in London. Kitty I love you as you've always known
dearly and for every good reason. Perhaps my luck will turn and
we will have a really good future." In stark contrast with his previous statements, he wrote,

> I will I think find a generous acceptance here and not the
> grudging patronage I've had in England. Think of me darling as al-
> ways loving you and thinking of you. We were made for each other
> body and soul and nothing less than that will satisfy us. Be patient
> darling...I have great hope, but sustain me with a letter that will
> make me feel you are really mine I will believe you and only you...
> How glad I was to get your letter and to hear you can get yourself
> a few things you like. I'm happy to think that my money can get
> you little things that you want. I wish I had given you one or two
> pieces particularly the Meum mask which I think you like...

Kathleen had written to him from Herefordshire, to which
he wired a reply on January 8, 1928, "WILL COUNT EACH
OF THE 18 DAYS OFF WITH JOY BLACKHALL LET-
TER CAME WAIT FOR ME." A letter followed. "Sweetheart.
I have your unhappy letter written from Blackhall and it must
seem more and more tiresome to you I can see that I ask you
to wait...How I want to be with you and more than 'play' with
you. I cannot imagine two persons who can be more with each
other in spirit than you and I. Then I can imagine you saying
why don't you hurry and come back or why did you go away at
all. Will you let these considerations influence you?"

Letters flurried back and forth across the wide ocean: "I have two letters from you one in which you fairly accuse me of caring for nothing but money and thinking of nothing but that," wrote Epstein in January.

My gains I have here are nothing to boast of. I've had a bad time been badly presented and as I am new to everything American have not known quite how to deal with the situation here. I've been unhappy and without you I cannot live and you know it. Now that my going from here is determined I feel for the first time easier in my mind. Your words of love fill me with joy. Our bitterness I know comes from too intense and unsatisfied love. I have only felt once before in my life as wretched and chained up as I have here and that was when I was in the army. As things are now I feel I am escaping from a life I don't want to lead to one I do. I went again and asked for your letters at 565 5th Avenue [the offices of American Express] and got there a letter from London and three cables and how happy I was to get your cable about my coming back and how it was the "best news." How wonderful sweetheart to be with you again. I have only lived with you and our natures are so bound to each other that nothing I feel can separate us. Change won't come from me.

He wrote again at the end of January.

How despondent a note from you I have and I can only echo you for this long wait gets worse and worse. I think forward and always the happiness that seems so far off though certain mocks me. We think and feel alike and write and write our thoughts that are the same. With you I have every joy and every happiness. I have had such delite [*sic*] as I'd ever dreamed possible and if I could only hold you again sweetheart I'd want nothing more. It is useless almost trying to bear up: we must be together again as we have been and that will give us any kind of contentment. You write that Helen has taken a place of her own: is that so she doesn't try to make you have in her lover to the extent of his becoming a habitive of No 13? How stupidly I begin to think but I imagine our place for ourselves alone and you would rather have it that way wouldn't you?...Sweetheart I've just been and got your letters and how happy and cheerful you sound and I'm glad. To look forward to seeing you every day the thought of that makes me happy. Never mind about the clothes except that I often see things here I'd like to send you but most likely the duty on them would be as much as they cost. I'll get things for you in London if I have any money left...heavens Kitty to be with you again what joy. Why didn't I meet you when I was a boy and poor as could be and we would have begun our life together and there would have been none of these complications that we have now and we would have been perfectly happy...

At last the day of his sailing arrived. The thought of seeing Kathleen filled him with tender anticipation.

You may ask if I look forward to seeing you; by the time you get this I will be most joyfully on my way to you...I know where I am happy more happy than I can tell and that it is for me only is my great pride with you and I am glad to be alive and I worship with you. Never mind about the splendours of New York: my ideas of splendours are quite apart from such things as shop windows and stupid social doings which seem to fill London papers as well as New York ones. We have known happiness together with little or no money at all. I experience everything with you and that is what I look forward to. Why cannot I be your lover and why must you think of fine clothes for me. You are beautiful and I've shown it in studies of you and I've seen such loveliness in you Kitty as would take me a life time to repro- duce ... My plans for work are not very clear yet. I first want to get back and be with you. That is the great thing and the thought of it fills my mind. I dream of things: to mention what perhaps may not be so impossible of realisation of being with you for lengths of time so that the swift happiness will have even more reality ... I anticipate the shirt you are sewing for me and as you have shirts of mine you ought to know the right size ... to be with my lovely girl again: the thought makes me impatient. Time and distance are not for us two Kitty ... How I would like to go away somewhere to see the fine things with you to Toledo and Madrid or the great Grünewald altar at Colmar.* Or to work on some- thing that was more than a portrait. To do a portrait seems my chief use to the world. May this week and the following go quickly. If I get on the Friday night [train] I'll be with you the fol- lowing day. Saturday night may it come with its joy ...

Epstein came back to England alone, sailing aboard the *Aquitania*, where he dined in the company of Lord Rother- mere, the millionaire Charles Schwab, and P. G. Wodehouse, listening with bemusement while they discussed stocks and

---

* Kathleen and Epstein did see the Grünewald altar together, but not until the end of the 1950s.

shares. Mrs. Epstein and Peggy Jean followed shortly after. No sooner had Mrs. Epstein arrived than she began house hunting. Leaving Bloomsbury behind, she settled on a tall house in Kensington, 18 Hyde Park Gate, which had a huge studio at the back and stood in a quiet cul-de-sac. The writer Enid Bagnold* lived on the other side of the narrow road, and the Stephen sisters, who were to become Virginia Woolf and Vanessa Bell, had grown up on the same street. Robert Baden-Powell, founder of the Boy Scouts, lived at the park end of the road, and when Winston Churchill ceased to be prime minister, he set up home opposite, complete with a policeman permanently at watch on the doorstep (a guard afforded to all former prime ministers). Kathleen was now on the other side of London.

But not for long. Epstein wanted Kathleen to move west, too. "You are right about moving away from near Brunswick Square," he wrote her. "I would rather you did and if you could find a flat ... somewhere quite near me I think we could then see more of each other ... Also if you live in close proximity to your family I know I will be told all sorts of stories about it and I'd rather avoid all that." Epstein had never liked it when Kathleen's sisters, first Mary and then Helen, invited their admirers to stay at 13 Regent Square, and any talk of men coming to and going from Kathleen's address aroused his jealousy. Nevertheless, he was fond of Kathleen's sisters and they of him, especially Helen, who greatly loved and admired him: "He was the most vital person I have ever known," she said. For his part he used to grumble with

---

* Best remembered for writing *National Velvet*, the film of which launched the career of the child star Elizabeth Taylor.

playful exasperation that "the Garmans aren't a family: they're a tribe." So he was pleased when Kathleen followed in his direction, taking the top three floors of 272 King's Road, a house almost opposite Oakley Street.*

Always referred to as two-seven-two, this house was to be Kathleen's home and a roof over the head of Garmans old and young for many years to come. It was one of a terrace of shabby but elegant houses with graceful wrought-iron balconies, giving on to a long garden at the back. Kathleen furnished it simply: the Egyptian mask of Nefertiti which Epstein gave her was on the piano, there were some of her favorite lilies in a slender vase, and pictures, as ever, propped against the walls.

Like her mother, Kathleen serenely avoided housework. She never took to dusting and polishing, and she is remembered doing the washing up with her coat on, as if to escape it as soon as she could. "I never saw my mother in an apron," says Kitty. "She didn't even know what oven gloves were for." Although Kathleen was rather grand in manner, especially when she became more prosperous in later life, her unwillingness to do household chores was not an affectation. Any middle-class young woman such as herself at that time would have grown up expecting other people to cook, launder, iron, dust, sweep, polish, fetch, carry, and lay fires. A young woman could leave her father's home and enter her husband's without ever having bought half a pound of butter or a box of matches for herself. In this respect, the only difference between a doctor's daughter and a duke's was that the

* Now demolished. A modern fire station stands on the site.

former would probably have brushed her own hair and buttoned her own bodice. As the aristocratic Nancy Mitford wrote in *The Pursuit of Love*, "I think housework is far more tiring and frightening than hunting is, no comparison, and yet after hunting we had eggs for tea and were made to rest for hours, but after housework people expected one to go on just as if nothing special had happened." The Garman girls had their minds on loftier things, anyway: art, friendship, music, books, love. Several of them did become expert gardeners and cooks, but none of the Garmans liked to look as if they were trying too hard, and although Kathleen cooked well, it was her skill at making a meal seem as if it had just happened by chance that people remembered.

Lorna's elder son, Michael Wishart, recalled his aunt and her house in his autobiography *High Diver*.

I was very lonely. I sometimes went to see my mother's sister Kathleen. She played the piano to me. I asked to be given lessons by her. I was an exemplary failure, but I loved my lessons because I idolised my aunt. She was poor. She lived in a beautiful Georgian house in the King's Road, Chelsea, with my cousins Theodore and Esther. I found them and their house very romantic. They lived surrounded by bronzes, drawings and gouaches by Modigliani and Epstein, amid the marvellous Benin heads and furniture which Epstein and Picasso had collected in their impoverished student days, and which had such an influence through Picasso on Cubism and through Epstein on Vorticism ... Chelsea in those pre-war days had almost nothing in common with the district as it is today. It was something like a tranquil market town: a real artists' quarter, and more romantic than the rest of London. The King's Road was a quiet residential street, enlivened here and there by chemists' shops displaying large coloured bottles which gave an air of stained glass windows in old churches. Family grocers and a few small inexpensive restaurants all added to the homely domestic atmosphere of an untroubled life-style which exists nowhere today.

Wishart recalled local personalities. "Augustus John, every inch the artist, like a character out of *Trilby* or Murger's *La Bohème*; Brenda Dean Paul with her silver mane and slacks, scarlet mouth and man's overcoat to match, tottering with her friend Miss Baird to refill her hypodermic syringe; my aunt Kathleen, hastening to a concert or a tryst, laden with Beethoven scores; Quentin Crisp, pioneer of Gay Lib, mincing proudly to and from the cosmetic counter at the nearest and dearest chemist … It was not very rare to see the curly-haired boy Dylan Thomas emerging from a pub, backwards, very fast and seldom from choice …" The Chelsea Arts Club was only a few minutes' walk away in Old Church Street, as was the Queen's Elm pub, which was later to become the unofficial London residence of Laurie Lee.

On the first floor of 272 was a double drawing room, where a Cameroonian stool of carved men and leopards was used as a stand for a fruit bowl and for the flowers that Epstein brought; the stool is now in the Garman Ryan Collection. Epstein was among the first serious collectors of what was then known as tribal art, and Kathleen was a pioneer in displaying such pieces alongside European paintings and furniture.

"The whole atmosphere was of high Bohemia," Kitty recalled. "There was only one key to the house on a perilous string which hung inside the letterbox. Student friends, old musicians and young artists from the nearby studio would simply fish for the key and walk in. (But no-one was allowed in when my father came between 6pm and 7pm to visit my mother.) He sometimes took us all out to the Isolabella restaurant or to the Ballets Russes de Monte Carlo and to Sadler's Wells. Other painters came, certainly Matthew Smith, who was my father's oldest friend..."

As Kitty and Esther grew up and came to stay with their mother in London, Kathleen took them to see Vivien Leigh as Titania in *A Midsummer Night's Dream*, to Chekhov plays, and to the popular classical music concerts, held at the Albert Hall, known as the Proms. There was a visit to Swinburne's house and to the Victoria and Albert Museum and to lunchtime concerts at the National Gallery, which were free. She even took them rowing in a boat on the Serpentine. Every now and again she'd say, "Come on, chicks—we're going to Richmond," and there would be a picnic in the park there, or by the river. Kathleen always enjoyed what she and her siblings called "a jaunt."

Kathleen had a genius for getting people to do things for her. She never learned to drive because she never needed to. (Epstein would say to her, "Kathleen, if you were to learn to drive, I know your whole face would change." According to Kitty, "She did indeed have a very idiosyncratic expression, being both uplifted and abstracted at the same time, as though she were far away in a world of her own.") She somehow always managed to procure a car and a willing chauffeur from among her large stable of friends, so there was generally someone to escort her on essential visits, or simply to take her and her entourage on picnics. There was a kind of reward for these unpaid helpers in the way Kathleen could magic things out of hats, making everything seem fun, special. She'd find a place for lunch, a lovely wild-flower meadow beside a river. Delicious and rare French cheese and wine would appear from her basket.

Picnics, undertaken in all seasons, were a Garman staple. The combination of spontaneity, informality, and being out of doors was very much to their liking. That you didn't have to do any washing up was another advantage, and since most of the family were quite hard up, picnics were also an inexpensive and festive way of feeding people. These moveable feasts provided an easy means of accommodating large numbers and several genera-tions. For birthday picnics there would be roasted chicken and champagne. On winter picnics the focus would be on making a huge fire to sit around.

The Garmans would never have eaten in a parked car or by the side of the road: the whole point was to find the most un-spoiled and beautiful spot, the best view. The top of Bulbarrow in Dorset, from where you could see five counties, was a favorite. After Mrs. Garman moved to Sussex, she would take a taxi as high up the Downs as the road would reach and then walk the rest of the way, to reach a point where she could see the sea. Kathleen and Lorna preferred woods, but the others loved get-ting up into the hills, which led their children to refer to all pic-nics as windy-picnics. There were picnics in ancient country graveyards, at Manorbier Castle in Wales, and on the shingly beaches of the south coast. For Roy and Mary Campbell, who lived in the favorable Mediterranean climate, most meals could be taken outside, shaded by vines for wine-drenched lunches and lit by the stars in the balmy evenings, when Roy would make his own bouillabaisse and sing a husky, heartfelt rendering of "The Bonnie Earl of Murray" into the rose-scented night.

As the children of Kathleen's brothers and sisters came of age, they, too, would pitch up at 272. For those down on their luck, Kathleen tried to find a way to help: an impoverished young Indian woman was brought in to give drawing lessons to the children; an out-of-work Russian ballerina taught steps to Esther and Kitty. A gaggle of young Irish people who had just arrived in London were walking past the house one day. Kathleen, who happened to be at the window, liked the look of them, so she called out a greeting and they came in and made friends. Two-seven-two was like that, always: open house.

People, especially the young, were attracted to the Garmans. They all cultivated an atmosphere of familiarity and ease, and they were generous and teasing. Each had his or her own particular set of nicknames, mimicry, and jokes, which made people feel as if they belonged to a select club where ordinary values and judgments were suspended. The Garmans had more intimates than acquaintances, and, in truth, some of their close friends were almost acolytes. Kathleen and Lorna, especially, were surrounded by admirers. It would have been easy for them, with their beauty and their artistic connections, to become part of the Bloomsbury circle, or to be among Evelyn Waugh's Bright Young Things, or to join the raffish Fitzrovians. They knew people in all these circles. Their poise could have gained them admittance into grand society, too, but standing on ceremony never appealed to them. They were oddly shy. Kitty says of Kathleen,

She was certainly (like all the Garmans) afraid of "Society" and Social life. She was so strange and diverse that she didn't come under any category—Vamp, Arty Lady, Hostess etc. She was not of the Bloomsbury Group *at all*, and despite all her knowledge of music and literature, entirely unsophisticated... she sort of picked up people she had met along the wayside, or at a concert at the Albert Hall or Wigmore. She had a host of friends who worshipped her, young students, impoverished out of work ballet dancers, old refugee musicologists fleeing from the Nazis; a whole Italian family, aristocratic Jews who had come to settle in Kensington because of Mussolini... Odd as it may seem, there was a certain innocence and naivety about her, as there was in her sisters, even in Douglas. I think she never went to a night-club or set foot in a pub, or went to Wimbledon or the races... I think they inherited this kind of unworldliness from both my grandparents—Walter and Marjorie.

Their children remember their mothers after their dancing days. In youth, Kathleen had certainly been to nightclubs, frequenting the Café Royal and the Gargoyle and the Cave of the Golden Calf, a tiny subterranean joint opened by August Strindberg's second wife, which had been decorated by Epstein and Wyndham Lewis. Both the youngest Garmans, Lorna and Ruth, thoroughly enjoyed dressing up and dancing, as they had practiced doing when they were girls, on the lawn at Black Hall. Ruth would sometimes meet Lorna in London for a night out. The two egged each other on, and both of them had a brief fling with Leslie Hutchinson (known as Hutch), the fashionable Caribbean cabaret artist, who played the piano at the Café de Paris and Chez Victor and later at Quaglino's.

Lorna, who had her first child at seventeen, would leave her young family to drive off alone in her huge, expensive car for a night out, all the way from Arundel in Sussex to the West End of London, dressed in ultramarine sequins to match her eyes. Her son Michael was later to write in his memoir *High Diver* about her nocturnal leave-taking, with self-conscious echoes of Proust: "My mother leans over me. Dressed for dancing in clinging sequins...she resembles a sophisticated mermaid...after mother's departure, I bury my face into the cool pillow where the scent of *Fleurs de Rocaille* lingers. I listen for the familiar purr of the chocolate brown Bentley, crunching on the gravel driveway. My mother, always alone, is speeding through the darkling hawthorn heading for nightclubs which assume, in my half-asleep loneliness, vague Xanadus of Kubla Khan." Lorna's spell was so powerful that even her own son fell under it.

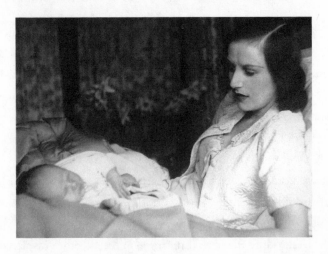

Lorna with her firstborn, Michael, circa 1928.

Chapter Seven

MARTIGUES

When they moved to France in 1929, Mary and Roy Campbell rented an old stone farmhouse called Tour de Vallier, close to Martigues. The house stood on a low hill surrounded by fragrant pines, with Cézanne's Mont Sainte-Victoire to the east and Van Gogh's Saint-Rémy in the Alpilles to the north. Every morning they took their coffee on a terrace beneath the umbrella pines, overlooking the vast lake, the Étang de Berre, which glinted through the trees. They rose late, Campbell having worked long into the night, writing the poems which would in time be published as the satirical *Georgiad* and the lyrical *Adamastor*. With her husband thus occupied in the attic, Mary made herself a painting studio from the top floor of a little cottage that came with the property, just beyond the well from which they drew their water. The studio was furnished simply with only a low couch covered with a beautiful piece of Indian cloth, and an easel. Here she painted portraits of her

husband and their daughters, as well as landscapes and seascapes. A portrait of Tess was praised by Augustus John, who encouraged Mary to continue painting.

It was with Augustus John, in 1927, that the Campbells had first seen Martigues. As early as 1910, he had become entranced by the place, and for eighteen years afterward he kept a house, the Villa Sainte-Anne, on the edge of the town, overlooking the lake. Here John's lover, Dorelia, and their children would row in their little boat, while John took

Mary in the south of France during the early 1930s.

the donkey off on sketching trips in the surrounding country. "Martigues is like some rustic mistress one is always on the point of leaving," wrote John, "but who looks so lovely at the last moment that one falls back into her arms." John also made a study of local Gypsy lore and occasionally took to the streets of Marseille, going through the narrow streets of the port in a huge cloak, to visit the "fine assortment of Mediterranean whores" there.

During their first year at Martigues, the Campbells saw a lot of the Johns. Aldous Huxley, who lived not far away at Sanary, was another visitor, coming to stay for regular weekends. Once they all went swimming from the beach at Sanary, Huxley taking to the water in a small blue rubber boat. Afterward, he helped Mary to dress, feeling for the complicated buttons of her shirt with his long fingers, since he was too shortsighted to see them. The young Sybille Bedford, who later became Huxley's biographer, was introduced to him by the Campbells. She remembered that Roy in those days drank red wine almost continuously, starting at breakfast. Nancy Cunard arrived in Martigues, causing a stir with the Campbell daughters by being taken about by a very big and jovial black chauffeur whom she called Darling (this was her African-American lover, the pianist Henry Crowder). The children were fascinated by her dangerous aura of scent and dry martinis, and her legendary bracelets, worn all the way up her sinewy arms. They remembered her as resembling a beautiful but overstimulated snake. The painter Tristram Hillier also came to stay with the Campbells and included their sheepdog (the gift of a fishing friend of Campbell's who was innocent of his hatred of dogs) in a family portrait. Mary's sister Lorna came for a visit with her husband, but they annoyed Campbell by complaining about the plumbing instead of appreciating the beauty of their surroundings. The Irish writer Liam O'Flaherty arrived with his daughter, creating a lasting impression on the Campbells' little daughter Anna, who, even sixty years later, remembered him as the best-looking man she had ever seen, and one of the wildest.

Hart Crane, who had been staying in Marseille, arrived at Tour de Vallier for a fortnight, puzzling his host by composing poetry straight onto his noisy typewriter. To Campbell, the silence after midnight was the optimal time for writing, and the pen was almost sacred (Laurie Lee was to describe him as one of the last pre-technocratic, big-action poets). For his part, Crane very much admired what he called Campbell's wonderful, splendid poetry. But the two of them did not only write and discuss their work. Campbell had to restrain Crane from jumping off a bridge in a fit of alcoholic depression and was obliged to intervene when he made an unwelcome advance on a local sailor. "He is a good poet...but he went off his rocker completely," Campbell told a friend. "We had to get him to go away but we parted good friends...we found him sitting at 12 at night in the middle of the road with his typewriter on one side of him and his portmanteau on the other, crying like a baby." Campbell's account of the incident took on a brasher tone when he relayed it to Wyndham Lewis, now describing how Crane had started banging on the table until Campbell threw a bucket of water over him to sober him up. Campbell had kept him up till two in the morning, going over everything Crane had said, until the visiting poet began to weep. In the end, the wretched Crane said he would come over to Europe with a brigade of Americans to shoot every European. "His chief quarrel against Europe it seems was that the sailors and fishermen were not pederasts. I chucked him out of the house when he became so truculent," boasted Campbell, whose machismo was always to the fore in his dealings with Wyndham Lewis.

Crane went on to Paris, where his friend Harry Crosby tried to get some food into him and curtail his drinking. "Hart Crane back from Marseilles where he had slept with thirty sailors and he began again to drink Cutty Sark," noted Crosby. Yet Crane himself had happy memories of his southern visit. "I have come to love Provence," he wrote, "the wonderful Cezannesque light (you see him everywhere here) and the latinity of the people. Arabs, Negroes, Greeks, and the Italian and Spanish mixtures. When I come to France again I'll sail direct for Marseille—and it's certainly my intention to come again!"

Another Martigues visitor was the writer Kathleen Hewitt. Now obscure, she wrote a series of detective novels, other novels, drama, short stories for the London *Evening Standard*, as well as journalism. A satirical novel called *Decoration*—dedicated to Mary and Roy—became a best seller. Her autobiography, *The Only Paradise*, recalls her first meeting with the Campbells, during the *Voorslag* magazine days, in South Africa. Then, she had been struck by Mary's dramatic beauty. They all met again in London during the late 1920s. Hewitt describes a visit to Martigues, remembering the sweet-scented herbs, olive groves, vineyards, and pines which grew all around the house. She remembers that they had to walk a mile to fetch buckets of water. By day, the sun blazed gloriously, and in the evenings they cooked meals over a fire of sticks in an open hearth and drank the local vin rosé tipped from a great wickered flagon. Campbell made wonderful mussel stews that they ate from bowls with yards of crisp French bread, and they all talked a lot, arguing, laughing, boasting, and contradicting shamelessly. Campbell would reduce Mary and Kathleen Hewitt to tears with his violent reenactions of comic incidents, including his experiences in the bullring. Sometimes she and Mary would go to Marseille and return with their shopping baskets full of knickknacks, and sometimes the Campbells invited friends up to the house for evening parties that went on long into the night. "It was all grand fun, but I was almost glad to get back to London for a rest cure," she concluded.

Roy had tremendous energy. He not only wrote and entertained friends, he took to local people and their customs with zeal. He bought a boat, and then another; he fished the lake and the sea; he jousted and fought bulls. Like Hemingway, he was an extremely sensitive man who found a kind of resolution to his own inner conflicts in the archetypal wrestling of men and beasts, fishing and bullfighting in particular. For Campbell, whose sensibility was so greatly influenced by the classical, the parallels between ancient Mithraic legend and the tournaments of the ancient Greeks with these Provençal sports made them all the more compelling. Wyndham Lewis observed as much.

Campbell paid a price for liking Lewis, who was reviled by many: Augustus John and Aldous Huxley dropped the Campbells because they could not stand Lewis. During the years of their friendship, Lewis could hardly write a novel without putting a barely disguised Campbell in it. In *The Apes of God*, Roy appears as Zulu Blades. And in *Snooty Baronet*, published in 1932, Lewis describes his visit to Martigues in 1930. Here, Campbell appears as Rob McPhail, who has gone to live in the south of France with his wife and there taken up bullfighting.

Lewis also draws a vivid portrait of Mary (she is called Laura in the book) and her sister Helen, who was staying with the Campbells at the time. The hero travels to the south of France, where he hopes to persuade McPhail to join him on a trip to Persia in order to explore the cult of Mithras. It is evident that McPhail is not going to join this expedition. Instead they all sit about in cafés and bars, drinking and talking. Mutual enmity is touched upon: at the word "Bloomsbury" both men spit, one to the left, the other to the right. The narrator notes the "delicately chiselled" face of Rob's wife. And the Garman eyes were not lost on Lewis. "I glanced into the bashful black violet of her eyes," he writes, later adding that Rob is rewarded by a glance from his wife with, "A velvet sparkle of softly shining eye."

Wyndham Lewis noted Mary's unfathomable calm in a scene where her husband has been badly injured in the bullring. "Laura McPhail looked at me but said nothing, and my confusion turned into a professional curiosity as I gazed for a moment into the stoic depths behind her glances—whose eyes, monotonous, immobile, violet-velvet had that *pool-quality*—which it would take more than a dead hero (struck down in a fifth-rate bull-fight ...) to ruffle." Lewis detected that although Mary could be whimsical, there was something hard in her, too. Following the McPhails back to their house, Rob unconscious on a stretcher, the narrator waits to see if he can be of any assistance: "I looked up as Laura's sister [Helen], a big blonde woman, with pale blue eyes, came out of the sickroom, with a sort of staring, strained and stormy look. This woman's lack of *fatalism* appeared to me at the moment I must confess as affected as her sister's mastery of same. These people can feel nothing, I thought, that is about the size of it. They can only mimic feeling, or control feeling."

Another character portrayed in *Snooty Baronet* is a local fisherman whom Rob has befriended, who is described as fawn-colored, like a lion. This was Marius Polge, known to all as Grandpère, despite his youth and beauty. A visiting artist, Bateson Mason, R.A., was so struck with Polge that he painted a portrait of him on the spot. His fair coloring, rare in the Mediterranean, derived from Norwegian ancestry on his father's side. Grandpère was very popular locally. He became Roy's greatest friend, and the two bought a boat together and dived for mussels from it, as well as fishing with nets. When Helen came to stay, Mary and Roy took her to a dance. Many years later, she recounted how she had always been nicknamed Fatty and no one had ever asked her to dance at parties. Not until that summer at Martigues, when this handsome blond fisherman approached her, smiling, his hand outstretched. Mary, who was sitting beside her, made to stand as he came over, but Helen was at once on her feet and took to the floor with him. After that, he came to visit her at Tour de Vallier, standing with his foot possessively on the rung of her chair. One day, he brought her a big fish wrapped in newspaper: "pour vous." They married in 1930, and their daughter, Katherine, was born the next year.

At first, Helen lived at her mother-in-law's house in Martigues, but the punitive regime of floor scrubbing and ham curing that she was put to soon had her finding a house of their own, as perhaps it was intended to. There were original Epstein drawings on the walls of the Polges' little villa, unframed and darkened by smoke from the wood fire. A very able typist, Helen was always the least work-shy of all the Garman sisters, and in time she got a job at one of the oil companies on the far side of the Étang de Berre, to which she would cycle every day. (That there was any such industry near Martigues was never mentioned by the romantic Campbells, who turned a blind eye to anything so ugly.)

As ever, the Campbell girls were left very much to their own devices. Rising early, before their parents, they would be given breakfast by a maid called Sérafine, who served them large bowls of café au lait and round slices of bread and butter. The children loved Sérafine, who was plump and unruffled and sat doing crochet in the kitchen for hours on end instead of keeping an eye on them. After breakfast, they would go out to play, exploring in the woods, running along shady paths which were scented with sage, thyme, and rosemary. They would go searching for wild asparagus and mushrooms. Often they spent whole days at the lake, sometimes picnicking with their mother while Campbell and Grandpère fished on the still waters. When their parents and Helen went out for an evening, they would roll up their small nightgowns and walk solemnly to a neighboring farm to spend the night.

Sometimes Campbell would take them with him when he went to Martigues to shop. First Mary would flutter a shopping list down to the terrace from her bedroom window, and then they would set off to buy fish and saffron, fruit and vegetables and wine. Having filled their baskets, they would stop at a favorite café for Roy to take a glass of red wine, while the girls had fizzy lemonade, which they called pins and needles. While their father was talking to friends in the bar, they would play in the shade of the plane trees around the fountain in the square. On other occasions, Roy took the girls fishing with him, setting off before dawn to catch mussels and sea urchins. "Both my parents were continually seeing beauty and pointing it out to each other and their children. It was in this way that they taught us to see life and it has certainly enriched our existence to an enormous degree," Anna was to remember. "We learnt to distinguish between beauty and trash." Such an appetite for life seems to have been the Campbells' greatest gift to their children. They were taught to look at the stars, to enjoy Greek and Roman mythology, French and English literature.

Roy was an affectionate father, but Mary tended to alternate between extreme inattention and sudden strictness. She had her mind on other things. In the first months at Martigues, she still wrote regularly to Vita Sackville-West. "Why do you write less and less often?" she demanded. "I fear it is indifference, Vita, or forgetfulness. I get so unhappy, so miserable—Vita it's unkind of you." Gradually the letters petered out altogether, and Mary turned her attention away from Vita. By now in her early thirties, Mary was still beautiful and as unconventional as ever. Although the fashion of the time was for midcalf-length dresses and bobbed hair, Mary would wear floor-length skirts and rope-soled espadrilles, with her thick hair falling to her shoulders. Sometimes she wore trousers which was, in the early 1930s, still very unusual for a woman. White was her favorite color of dress, offsetting her dark beauty. Roy bought her a fur coat from the proceeds of his latest book of poems, *Adamastor*. He loved Mary to dress in his favorite scarlet, and his taste for the theatrical also had them both in wide brimmed hats. She also favored rather Spanish-looking clothes: fringed shawls and capacious skirts and combs restraining her hair. Evenings at Tour de Vallier would draw to a close with Mary, thus attired, strumming her guitar and singing folk songs in her low contralto voice.

She had her admirers during the years in Provence. An old flame, a poet called Lazarus, who had taken her to Paris before she met Campbell, turned up to see her. Another visitor was Jeanne Hewitt, the friend of her London days. The marriage to Mary's brother Douglas was over by the time Jeanne came to France in 1932, bringing her sister Lisa with her. It seems that Mary and Jeanne became lovers at this time, a brief romance which Mary's letters (and Campbell's two biographers) attest to, although Jeanne's daughter remains convinced that they were never more than friends. Jeanne was not the only person to fall under the spell of one Garman after another. The Garmans' paternal grandmother had been much admired by their great-uncle, and when she married his brother instead of him, he promptly married her aunt. Among their own generation, too, more than one lover would be passed around.

It was during that same visit that Campbell began an affair with Jeanne's sister Lisa, who had once captivated young Mavin Garman and was, like her sister, a considerable beauty. The Campbells lived, briefly, in a happy ménage à quatre with the Hewitt sisters, but Mary and Jeanne's affair was soon ended, while Camp-

Jeanne Hewitt Garman, circa 1932.

bell remained captivated by Lisa, much to the annoyance of his wife. Once Lisa had left, though, Campbell attempted to placate Mary: "I wrote a semi-love letter to Lisa," he admitted to Mary. " 'May Venus shower a 1000 blessings on her valiant little soldier' that was the worst I said but I also said I was glad and better that she had gone away (and I only sent her my *spare* love *left over from Mary*. I love only Mary hotly and fiercely.)" Mary later described both these affairs as "very light-hearted and frivolous," but evidently they caused some pique at the time.[*]

While their mother's mind was on such things, the children

---

[*] Jeanne met Roy Campbell again, in New York in 1955. Her second husband spotted him hunched in a doorway, clearly the worse for wear, and brought him home to be nursed back to health.

ran fairly wild. It may be that Mary, who had been brought up under a Victorian regime at Oakeswell, wanted greater freedom for her own children than she had known in her own childhood. She may have been influenced by Augustus John and Dorelia's very lax child-rearing. Or it may be that she was mimicking her own mother's vague, live-and-let-live attitude to mothering, although Mrs. Garman was never unkind or critical or sharp-tongued, as Mary sometimes was. Like her mother, she may have hoped that someone else would do the hard work, but unlike Mrs. Garman she had no Miss Thomas in residence to provide it, and Roy was most unlike the disciplinarian Walter. Perhaps Mary was simply too self-absorbed. Tess Campbell grew up to believe that her mother had wanted sons, not daughters, if indeed she wanted children at all.

When Sérafine left after two years, the children were even less cared for. They were given occasional lessons by their parents in literature, music, and natural history, but other than this they were left very much alone. Roaming about, they would call in at neighboring farms, where they were given treats to eat and drink: tiny cups of strong black coffee, piquant salt cod, and cherries preserved in the local eau de vie. At the time, Tess was only seven years old, Anna just three: it was not surprising that they should have got upset stomachs from these tidbits, but their mother was exasperated and blamed the girls for their ailment. It was the same story when they contracted head lice: Mary gave them a severe dressing down, as if it were their fault.*

* Roy translated Rimbaud's bizarre and blackly erotic poem "The Louse Catchers," perhaps inspired by this infestation.

Mary had such an air of authority that the children never questioned but that she was right about everything. And she could laugh at her own faults so that, buffed with ironic self-deprecation, they became almost virtues. Lorna was the same: feeling guilty was simply not in the Garman makeup. Thus Mary did not blame herself when little Tess contracted severe sunstroke and raved deliriously for green fields and lemonade, nor when Anna got impetigo. Almost seventy years later, Anna Campbell still recalled the pustules and crusty sore caused by this skin complaint with horror, like "a heavy brown toad hanging from my upper lip." Far from treating her child, Mary went off on one of her sojourns to London, leaving Roy in charge. Anna always wondered whether her mother went to England in order to get some special cream for this sore, or whether she simply thought that it would do her husband good to have to cope with the children. It seems most unlikely that there was no ointment for impetigo in France.

Every six months or so, Mary would take the children with her to England. They would be left with her mother while she saw friends in London. The girls loved their visits to Black Hall, where there were cows and pigs and cart horses and barns full of hay and endless English puddings cooked by Miss Thomas. Regular meals were something of a rarity in their young lives. Mary's absences made Roy as ardent as ever. "My girl, think about me—I long to have your sweet face lying on my chest," he wrote, "I long to kiss you and drink out of your lovely eyes...I adore you and I am in a rage of impatience to see you...Kisses everywhere—all over you: 200,000 of them...

My little baby girl dream about me—I love you and want to kiss you a thousand times."

Back at Tour de Vallier, Mary wrote,

> After being in London I got almost a shock (a pleasant one) to find how rich and stimulating this rough life is. The day after I arrived there was a great feast at our house. Five jousters fishermen invited themselves to a bouillabaisse which they brought caught and cooked themselves. They made a large fire in the open air and all helped with this sacred rite! One felt they were the remote descendants of the men the gods loved. I don't love them;—as soon as I had eaten and drunk with them I hid myself in my room,—but I must say I have more respect for them than many rooms full of the London Paris New York intelligentsia. I believe there is a quality—in this air and soil and sea—to be found no where else in the world.

The recipient of these thoughts was William Plomer, Roy's friend from South Africa. When Plomer arrived in England in 1928, Roy wrote several letters of introduction for him, and he had an open invitation (never taken up) to visit them in France. He corresponded with both the Campbells, although Mary wrote more often. Despite her love of her surroundings, a wistful note begins to sound in her letters. "I sometimes long for a day in London"; "if you can spare me the time write me a long letter sweetest William." It is evident from the correspondence that the Campbells still experienced sporadic money troubles, despite Roy's income from his father's estate. Mary thanks

Plomer for a loan (asked for by Roy, without her knowledge),
promising to repay it by the beginning of the following month.

Mary's letters begin to suggest that she is flirting with
Plomer. Although he was homosexual, it may be that his fastid-
iousness combined with her own greed for admiration blinded
her to the fact, or perhaps she imagined he was bisexual. When
in London, she writes to say that she wanted to see him before
she left, while the next letter begins without salutation, "I'm so
bored you are not here. Please ring me up as soon as you get in.
I am staying with Lorna and my number is Terminus 6043. If I
shouldn't be there leave a message whether you can meet me
tomorrow evening for dinner..." Back in Martigues, she re-
sponds to the book of his poems he has sent, ending, "Goodbye
life is as you say so complicated and more than that Mary."
Writing again to praise his poems, she concludes: "Think nothing more of any disagreeable thing I said to you and if I said it
clumsily forgive me." In December 1929, she wishes him season's greetings: "I shall give you a present when I come to London, either the 6th of January or the beginning of Feb and I am
looking forward to seeing you looking at me—one eye full of
paternal cynicism the other of uncritical love." Another letter is
dated May 1930. As she often does, Mary dispenses with formality, beginning instead, "William darling you're very unkind
to go off with other friends to romantic places like Greece and
put off coming to see us, thinking that we shall be satisfied with
continued assurances on paper that you will come at some future date. What is your hidden reason for not coming?"

But the friendship came to an abrupt halt after Mary had

again seen Plomer in London, in August 1933. There were Chinese whispers: William Plomer told Laurens van der Post, who told another South African friend, who in turn told Campbell that Mary had made a pass at Plomer in a London taxi. Mary denied it. Campbell was furious. He fired a letter off to Plomer; "knowing your general hatred of women and marriage in particular coupled with your nancydom it points to a very different sort of triangular predicament..." As was Roy's custom, he turned his rage into scathing satire, ridiculing Plomer in verse:

Pale crafty eyes beneath his ginger crop,
A fox's snout with spectacles on top . . .
In him the "friend" concealed the jealous "tante"
Who slandered women he could not supplant,
Whose faults he would invent and then reveal
On the pretext of trying to conceal.
He'd blurt a secret (none so sure as he)
By hiding it so hard no-one could see.

No more letters were exchanged. William Plomer went on to become a friend to Virginia Woolf, Harold Nicolson, and others of the Bloomsbury set, the very people whom Roy had derided in *The Georgiad.*

In 1932, the Campbells moved house. Figuerolles was bigger than Tour de Vallier, an old pink farmhouse farther from the town, with an enormous cedar tree in the middle of a rough lawn. Here Tess spent most of her time looking after rabbits, chickens, and a white goat which she called Blanchette. The

Campbells' improved circumstances, with money made from *Adamastor* and *The Georgiad*, meant that a tutor was hired for Anna and Tess. The first incumbent of the post was Marie-Louise, whom Anna remembered as looking like an overmade-up pug and who fell foul of Mary by flirting with Roy. She was soon replaced by the South African poet Uys Krige, who had been staying in Marseille. The children liked Krige, but he was incapable of disciplining them. Anna used to mimic him, much to his irritation. He got on well with Roy and became very friendly with Mary. "I was jealous of the time he and Mary spent together," said Anna. "She painted him in the nude (a wonderful painting, rather like a Botticelli) so that I feel, even though I did not know, that they were having an affair. It must have been a light-hearted intimacy as there were no tears or scenes." Mary had indeed taken Krige as a lover, and he began to spend less and less time on the children's lessons. "People were horrified by our lack of manners," says Anna. "We were never told how to sit at a table, or how to place our knives and forks after a meal was over or how important it was to change our knickers every so often." Mary spent hours painting, often out of doors, off on her own with a camp stool and easel and oils. The children and the latest maid would sometimes bring her a picnic lunch and a bottle of white wine.

The musician Cecil Gray came to spend the New Year at Figuerolles, and much to the delight of the little girls, there was a light scattering of snow that Christmas. Another presence in their lives was a sculptor called Freddy de Fromanville, a French count. Short and highly cultured, he would wiggle his little finger whenever he wanted to emphasize a point. He lived nearby but spent a lot of time with the Campbells, who came to regard him as one of the family. Mary's mode of dress troubled him, and he insisted on introducing her to a Martigues dressmaker who, for a time, succeeded in smartening her up. When he went away, he always brought presents back for the family: a first edition of Rabelais or a special bottle of vintage wine. Helen, too, became very close to Freddy and enjoyed talking about poetry and art with him, subjects with which her husband was less familiar. After Mary left Martigues at the end of 1933, Freddy was the only person Helen could really talk to. It seems that he was another person to become close to successive Garmans: "Both my aunt Helen and Mary told me that Freddy was besotted with them," Anna Campbell recalled, "so I suppose that when my parents left Martigues, he transferred his love for Mary onto Helen." Many years later, when Helen had just married for the second time, she and her Italian husband went to Paris. Having heard that Freddy—whom she had not seen for a long time, not since before the Second World War—was there, she dragged her husband across the city to where her old friend was living. When she saw that he was very hard up, living in an attic, she at once invited him to come and live with her. Whatever her husband thought of such a proposal, such spontaneous generosity was typical of Helen in particular and the Garmans in general.

The Campbells left Provence after only a year in their new house, having once again become short of money. This time, Campbell hurried to complete a volume of autobiography, *Bro-*

ken Record, to bail them out. In the meantime, he and his family hoped to find a more affordable life across the border in Spain. When a pet goat of Tess's broke through a neighbor's fence and destroyed a number of valuable young peach trees, the neighbor demanded compensation. Failing to get it, he sued the Campbells for a considerable sum of money and won. Typical of their love of drama was their flight from France. Roy couldn't pay, so the family fled in the dead of night, leaving everything but their clothes and books behind.

Chapter Eight

PEGGY AND THE PARTY

A fter his visit to Leningrad, Douglas Garman spent the late 1920s between London and Paris, supporting himself by teaching English when in France and writing book reviews while in England. In France he spent much of his time in cafés, smoking and scribbling into elegant French notebooks fragments of novels he would never complete. He also wrote poetry. There was a rumor that he had fallen in love with a Bolshevik woman in Russia and had stayed on with her for six months, after sending Jeanne back to England alone. His marriage was unraveling during these years, and it may have been to get away from this sadness that Douglas went to Brazil to stay with his brother, Mavin, where an injury to one eye left him with a slightly tilted smile.

In the spring of 1933, Douglas returned to London, where he began working for Ernest Wishart's publishing company. One day he received a note from his old friend John Holms. The two had been close in the mid-1920s, when the Garmans had borrowed the Holmses' London flat, where their daughter was born. Douglas had published a piece by Holms in the *Calendar*. Many thought that John Holms had it in him to be a great writer, but he drank too heavily to apply himself to work, and talked his ideas away into late nights and the ears of countless women. He was now living with the American heiress Peggy Guggenheim. It was through Peggy that Holms had become acquainted with Djuna Barnes, and he approached Douglas to see if he could help publish her novel *Ryder*. The book had already appeared in America, and Holms thought that Douglas could use his influence with Wishart & Company to bring out an edition in England.

The couple met to discuss the idea with Douglas at the Chandos pub in London, and it did not escape Peggy's notice that Douglas took an instant shine to her. For some reason, a moratorium on American banks had been declared, leaving Peggy temporarily penniless, and Douglas offered to lend her five pounds in cash. The sheer novelty of being offered money amused her. When he asked whether he might come to stay with her and Holms in Paris that Easter, she said yes. Peggy has been described as disconcertingly candid, a quality to which Douglas would not have been a stranger, having been used to the company of Lorna and Helen, who likewise spoke their minds.

Peggy Guggenheim had yet to become a patron of the visual arts. At the time when Douglas met her, she was in her mid-thirties and her focus still tended to the literary. She was quick-witted and immensely curious, and she looked to the men in her life to satisfy that intellectual curiosity. She was willful, utterly unconcerned with convention, funny and generous, intolerant and extremely kind. In 1922, she had been married to Laurence Vail, an artist and writer. They had two children, Pegeen and Sindbad. During her twenties, she had lived in Paris, where she met Hemingway and John Dos Passos and Hart Crane, Man Ray and Scott Fitzgerald. She gave considerable financial help to Djuna Barnes, which was tacitly acknowledged when Barnes's masterpiece, *Nightwood*, was dedicated to Peggy. Vail had been violent and often drunk; the marriage ended when Peggy fell in love with John Holms in 1928.[*]

She was surprised to find herself so taken with Douglas while she remained in love with John Holms. Describing Douglas's visit to Paris, she says simply, "When he came, I fell in love with him." But she liked that Douglas noticed little things about her, commenting on her clothes, something the intellectual and inebriate Holms never did. One day in Paris, while the maid was sick, Douglas found her sweeping the floor. He took the broom from her and insisted on cleaning the house, because, he said, she was doing it so badly. They soon found that they had an identical recurring dream, of encountering an unknown island with three smokestacks in the middle of the Atlantic Ocean. Such coincidences often characterize the beginning of a love affair, and the shared discovery of them can make an idle fancy take on the gravitas of destiny. When Douglas went back to England, the two did not meet for several months. "But I had a strange premonition that in one year I would be his mistress," she recalled.

Peggy tried to stop herself from contacting Douglas, only once inviting him to a demure tea at the house she and Holms

[*] Vail later married Jean Bakewell, another American of means, whose first husband had been Cyril Connolly.

were sharing, off Knightsbridge in Trevor Square. Douglas brought his daughter Deborah with him, and Peggy was struck by how much they resembled each other. "Garman was exceedingly handsome," she wrote. He "had brown hair and eyes, one of which was damaged, and a beautiful nose, slightly tilted. He was pale." One night, a few months later, Peggy got drunk and taunted John Holms about her desire for Douglas, admitting that she had written to him, suggesting that they should meet again, this time alone. According to her memoirs, Holms was livid and punished her by making her stand naked in front of the open window in December and throwing whiskey in her face. After Christmas, Peggy took Sindbad by rail to Zurich to return to his father, who was then living in Austria (her daughter lived with her, but the boy lived with his father). By the time she had traveled alone back to Victoria Station, bitter thoughts about John Holms had accumulated: she blamed him for all the years of agonizing separation from her son and swore "a terrible oath" that she never wanted to see him again.

Less than thirty-six hours later, John Holms was dead. The previous summer, Holms had fractured his wrist while riding on Dartmoor with Peggy. Despite being reset, the bones had never realigned correctly, and he had been advised to have a simple operation. This procedure had already been delayed once, when Holms had the flu. On the morning of the operation, he had a dreadful hangover, but Peggy had not liked to propose postponement for a second time. He died under the anesthetic.*

---

* Anesthesia was an imprecise science then. Peggy Guggenheim's former sister-in-law also died under anesthetic, in 1935.

Douglas wrote her a letter of condolence, offering to help in any way he could. Shortly afterward, she invited him for dinner and they became lovers. John Holms had been dead for only seven weeks, but if Douglas was attracted to Peggy's courage, he was soon to discover that it was only skin-deep, for she cried on his shoulder every day for a year and a half. Peggy felt guilty at beginning a new romance so soon after Holms's death and tried to keep Douglas hidden from her friends. When Antonia White—whose brilliant first novel, *Frost in May*, had been published the year before—came to stay at Peggy's Sussex house in 1934, she went into Peggy's bedroom one morning and saw a pair of Douglas's gray flannel pants on the bed. Peggy promptly put them on as if they were her own and wore them for the rest of the day, although they were several sizes too big for her. At breakfast, she and Douglas would arrive separately and greet each other with studied politeness.

Douglas fell deeply in love. He wrote poems to Peggy from their very first day together:

Doubting I lay, but you too brought
Tears from a world I had not shared,
Till in accord our bodies stirred
And broke the tyranny of thought
Finding again what treasure lies
In secret hushed between your thighs.
And stillness in the blood confirms
The age-old act we thus rehearse
An uncontracted universe
Swing[s] in the circuit of our arms,

For the wild gift between your thighs
Drove out the terror from my eyes.

Douglas quoted Shakespeare to Peggy, comparing her to Cleopatra. He took her to see his childhood home, Oakeswell Hall. They visited Helen and Mary in Martigues. Peggy noticed that Helen was not at all happy and felt that she wanted to leave her husband, but was held back by their mutual love for their child. Guggenheim noted that they lived a very simple life and Helen had to work hard at her job because his fishing did not bring in enough for them to live on. Douglas also took Peggy to Sussex to meet his mother, who now lived with Miss Thomas at Vine Cottage in South Harting, near Petersfield. Lorna and Wishart were not far away, at Binsted, near Arundel. Peggy was intrigued by the Garman sisters. "They were all extraordinary girls," she wrote. "In his youth Garman must have been overwhelmed by so many women, and preferred not to see most of them any more. However, I was fascinated by them and eventually managed to meet them all." She became particularly friendly with Lorna, whom she described as the most beautiful creature she had ever seen, with her enormous blue eyes, long lashes, and auburn hair. The friendship endured longer than Peggy's affair with Douglas, and Peggy continued to visit Lorna in the country. Lorna's second son, Luke, has an abiding memory of going with his mother to collect Peggy from the railway station and then stopping at a butcher's shop, at Peggy's insistence. She bought a piece of steak, which she proceeded to eat raw, straight out of its paper wrapping.

At first, Peggy was as captivated by Douglas, five years her junior, as he was by her. She found him a straightforward, honest person with a wonderful sense of humor, and a fine mimic. He disapproved of all snobbishness and chichi. He spoke beautiful English as well as excellent Russian, French, and Italian, and he was well-read and gave her books. During their first summer together, Peggy rented a house, Warblington Manor, in Sussex and invited Douglas's daughter, Deborah, and his niece Kitty to stay. The girls were about the same age as Pegeen, and they all got on well together, riding bikes and swimming in the sea at Hayling Island. Peggy loved Deborah. "She was just the opposite of any child I had ever known. She was so mature, calm, sensible, self-contained and well-behaved, and so little trouble. She was intellectual like her father and loved to read and to be read to. She had a wonderful influence over Pegeen... they got on marvelously and were soon like sisters. They used to dress up in a strange collection of old clothes and costumes we kept in a chest, and gave charades, plays and all sorts of performances." Deborah was fond of Peggy, too, and always remembered her sweetness. Peggy also liked Douglas's mother. "Mrs. Garman had about thirteen grandchildren. She lived entirely for them and for her own children, but she immediately accepted Pegeen and was very kind to us both."

At the end of the holidays, Peggy and Douglas took Pegeen and Sindbad to Dover, to catch the ferry. "They went alone together, right across Europe," she recorded, proudly. "Sindbad was eleven years old and Pegeen was nine. They were a wonderful little pair. Sindbad wore Tyrolean pants with a dagger thrust in his belt." Mrs. Garman was away, so Peggy and Douglas went to Vine Cottage to be alone together, the lease on Warblington Manor having expired. Douglas was editing a book for Wishart

& Company, leaving Peggy to lie in bed all day, reading Dosto-yevsky. When his work was finished, they drove to Wales for a holiday. They went for long walks and visited manor houses, staying in a little boarding house on a river, where they had a bedroom and sitting room of their own and wonderful five course meals, all for fifteen shillings a week. Peggy remembered that a mad family lived on the other side of the river, where they kept two daughters under the table, sewn up in bags.

Back at South Harting, Peggy looked around for a house to rent, so that Pegeen could go to school with Deborah and Kitty, both of whom were then lodging with their grandmother. She couldn't find anywhere to let, but there was an Elizabethan cottage for sale nearby, and Douglas suggested she buy it. Yew Tree Cottage was named for the five-hundred-year-old tree that grew in front of it. Peggy loved its exposed rafters and beams and the enormous fireplace, big enough for several people to sit in. Cows grazed in meadows only a few feet from the windows. The house had four bedrooms, two living rooms, one bathroom, a kitchen, and a larder. It was small by Peggy's standards, but the grounds compensated for that. There was an acre of land attached to the property, and over time Douglas made a beautiful flower garden, as well as building a tennis court and small swimming pool for the children and a patch of mown grass just for cricket. (Sindbad was especially keen on cricket, much to his mother's dismay. Peggy called it "that exceedingly dull game.") Despite all his hard work, Douglas did not yet move into Yew Tree Cottage, and Peggy became miserable every time he went away.

That Christmas, Douglas accompanied Peggy to Kitzbühel to visit Sindbad. Douglas had never skied before, but he was venturesome enough to have an accident on his very first day, breaking several fingers. He had to wear a stiff apparatus to pull his fingers back, a contraption that was a foot long and stuck out in front of him, preceding him wherever he went. He was uncomfortable enough, not least since he was vacationing with Peggy's former husband and his current wife, a writer named Kay Boyle, who was much despised by her predecessor. Douglas, however, had published a story of Kay's in the *Calendar of Modern Letters* and displeased Peggy by getting on well with her. Yet despite their prickly relations while on holiday, Peggy begged him to move into Yew Tree Cottage on their return and bring Deborah with him. Kay Boyle annoyed Peggy by questioning whether she was responsible enough to take on another child. "I told her Debbie was so responsible she would look after all of us," Peggy said. Kay and Peggy vied with each other about the children, but neither of them was, in truth, a very competent mother. Peggy was fairly neglectful, at least to her own daughter. To Deborah she was "incredibly sweet" and Douglas's niece Kitty never forgot that Peggy had given her an ocelot coat and beautiful Fortuny dresses.

Having succeeded in getting him to live with her, Peggy now persuaded Douglas to give up his job at Wishart & Company. But instead of devoting all his time to writing, he became involved with a magazine, *Left Review*. In the early days of his time at Yew Tree Cottage, Douglas belonged to the Labour Party and was also on the Portsmouth Peace Council. The *Left Review* put him in touch with a group of like-minded intellectuals who were members of the Communist Party. The editor of the *Review* was a poet, Montagu Slater, who was to write the libretto for Benjamin Britten's *Peter Grimes.* (Slater's sister later lived with Mavin Garman.) Then Douglas's old colleague and friend Edgell Rickword took over at the helm. Amabel Williams-Ellis was another young idealist who was involved with the magazine, along with her brother John Strachey, who later defected to the Labour Party. These were serious young people, widely read, highly cultured, and committed to political change. They hoped to print proletarian writing in the *Review.* "We didn't want to fill the pages with our own stuff," Edgell Rickword remembered. In 1934, Douglas, too, joined the Communist Party.

While Douglas was studying Marx in the shed he had constructed at the end of the garden, Peggy stayed in bed, reading Proust and shivering, turning the pages of her book with fur gloves on. Sometimes she took all the oil-burning heaters in the house into the bathroom and holed up there. Eventually, builders were called in to improve the fireplace in the sitting room. They were also employed to knock down some walls and close up the formal entrance door, creating one enormous sitting room. This was Douglas's idea. Like his sisters, he had a great flair for decoration, and he now set about remodeling the whole house. He installed a stove that burned day and night, and had special shelves installed for Peggy's vast collection of gramophone records. "He had wonderful taste and chose perfect furniture and glazed chintz hangings," noted Peggy. Outside the windows, he planted tobacco plants, whose sweet smell would waft into the room on summer evenings.

The children were happy; the house and grounds were agreeable; there was a flat in London for visits to the capital. But Peggy had not got over John Holms, and her old habit of taunting her lovers once again got the better of her. She began to drink too much and compared Douglas unfavorably to his predecessor, even going so far as to tell him that Holms had been bored by him, a cruel jibe, since the two men had been good friends. "After I had been with Garman about a year and a half I began to get the idea of running away from him," she wrote later, with deceptively childlike openness. "I tried it on various occasions, but he always got me back. I didn't want to live with him and I didn't want to live without him." Despite her reservations, the affair continued. "He still loved me very much, though I did everything to destroy it. I don't see how he could have endured me so long." Once she provoked him to such an extent that he slapped her in the face and then was so ashamed of himself that he burst into tears. In her diary she wrote: "fighting all day, f—— all night."

By now, Wishart had amalgamated his own company with Martin Lawrence, the official publisher to the Communist Party. The new firm, Lawrence & Wishart, founded in 1936, was to be as highly respected as it was radical. Douglas became involved, commissioning and editing. Llewelyn Powys and the surrealist writer Mary Butts published work with the company, as did the former miner Lewis Jones. Douglas worked especially closely with the latter, visiting him in Rhonda; Jones believed that Douglas was making a most valuable contribution to the urgent task of breaking down the division between workers and intellectuals. Peggy still preferred Douglas not to work, annoyed

by anything that distracted her lover from herself. Nevertheless, he joined the Welsh miners' Hunger March of 1936, to report it for the *Daily Worker*.*

In the summer of 1936, he and Peggy went to France, visiting Paris before going on to Martigues to see Helen. On their
way home they stopped again in Paris, where Douglas met the
translator who was working on a book about Mussolini, and
Gala, the flamboyant wife of Salvador Dalí. Peggy found her
handsome, but "too artificial to be sympathetic."

Another autumn at Yew Tree Cottage saw Peggy locked into
chilly domesticity. "I seemed to do nothing but look after
Pegeen and Debbie," she wrote. "They were always having colds
and flu. No wonder, as our house was so cold, and the temperature of the rooms so uneven. I seemed to spend weeks admin-
istering inhalations and medicines, taking temperatures and
reading to sick children." Far from putting her off mothering,
this constant nursing made her long for a child with Douglas,
but she failed to become pregnant.

They went sailing on the Norfolk Broads with Edgell Rickword and his wife, Jackie. Peggy thought Rickword was a good
poet and critic, but found him painfully introverted and very
shy unless he was drunk. (His shyness did not seem to hamper
his love life, and he numbered one of Epstein's most notorious
models, Betty May, among his lovers.) Though she enjoyed the
company, Peggy detested sailing, which made her restless, nervous, and irritable. She and Jackie, who were old friends, got off

* Douglas Garman's moving journal of the Welsh miners' march deserves to find a
publisher.

*Douglas in the south of France with Peggy Guggenheim
and Helen in the mid-1930s.*

the boat and left the men to putter on alone. They visited
Woodbridge, an eighteenth-century town famous for its sheepskin industry, where Peggy shopped with gusto. Yew Tree Cottage was gradually filled with natural sheepskin rugs, which she
ordered from Woodbridge by mail.

Not long after the Norfolk trip, Jeanne told Douglas that she
wanted to marry again.* At the time, it was necessary to prove
infidelity in order to obtain a divorce. Proof generally consisted

---

* In the 1930s, Jeanne married the actor Rollo Gamble. He later worked in television
as a director of photography, notably on the popular series *Crossroads*. He became a
director of one of the new independent broadcasting companies and then of Redif-
fusion, before the couple retired to the south coast. *Doctor Who* buffs may like to
know that Gamble played the part of Squire Winstanley in an episode called "The
Daemons," in 1971.

of a private detective photographing the guilty party in a compromising position with a corespondent.* Such cases sometimes appeared in the newspapers, and publicity of that kind was particularly tarnishing to the reputation of a woman. As a result, honor had it that the husband, whether adulterous or not, would go through with being photographed with a woman. Would-be divorcés often arranged to meet a photographer at a seaside hotel, where a professional corespondent, often a prostitute, was employed for the purpose. A double room would be booked, and tea for two ordered from room service. When the tea tray was brought to the room, the photographer would burst in and take suitably incriminating pictures.

Douglas obeyed the etiquette of the time when he refused to divorce Jeanne, insisting instead that she bring the divorce against him, so that she would not receive any unpleasant publicity. As Peggy noted, "He was a revolutionary at heart, but all his habits and tastes belonged to the class in which he was born." He was already in the paradoxical situation of being a practicing Marxist living with an heiress, and an American one at that. But he had integrity, which demanded that Jeanne's name should not be sullied. Peggy had reason to be annoyed, since her name was not to be so protected. "I protested violently because Mrs. Garman had left Garman long before I met him, and I considered this most unfair." This was not the first

* As an easy means of making themselves known to the photographer, male corespondents took to wearing two-colored shoes of the type popular with golfers. Such shoes came to be known as corespondent shoes and were considered too caddish for the feet of a gentleman.

time she had cohabited with a man, but she was quite put out by the proceedings. "The whole thing was very silly," she wrote. "We had to be found in a room together, Garman in a dressing gown and I in bed. A detective came from London early in the morning, so that the children would not know about it. After that he said he wanted to come again, but Garman said he would not go through it a second time, it must suffice. The divorce was granted...It read that Garman and I had sinned not once, but many times. I wondered how they knew. The only place this fact was ever recorded was my diary."

Perhaps Peggy was disappointed that Douglas did not ask her to marry him, once the divorce was granted. He had promised her that he would marry her if the king married Wallis Simpson. Their unweddedness could lead to social awkwardness. When a neighboring landowner shouted at Peggy for walking with her Sealyham terrier off the leash, Douglas was determined to write, demanding an apology. As it was, he was hampered by being unable to think how best to refer to her. Young Deborah suggested that he refer to her as his darling Peggy, and the letter went unwritten. Peggy was further embarrassed by an incident at Pegeen's school, regarding their different surnames. But with her low boredom threshold, she was only too grateful that their being unmarried prevented dull local gentry from calling at Yew Tree Cottage.

Nevertheless, Peggy needed the stimulus of company in order to enjoy herself. She felt that she put up with all the disadvantages of wedlock, without enjoying any of the advantages. "We had very few guests at Yew Tree Cottage and I led an extremely lonely life, getting more and more depressed. I was almost melancholic. The long winters with nothing to do but remain indoors, or take walks in the mud, or play tennis and freeze kept me inactive most of the time," she complained. "My only joy was reading." She reread her favorite book, *Anna Karenina*, and tackled *War and Peace, Wuthering Heights*, and the novels of Henry James. Douglas gave her the works of Defoe, and she also read Pepys's *Diary* and Countess Tolstoy's life of her husband. Of Tolstoy's wife she noted, "I started to ape her fiendishness and found myself behaving more and more like her."* In the evenings she read *Robinson Crusoe* and *The Last Days of Pompeii* aloud to the children. She began to drive into town several times a day, to buy things. Peggy always cheered up when she was shopping.

Peggy genuinely loved the countryside, at least when it was not too cold, and there were constant walks on the downs and through the bluebell woods that adjoined their garden. She relished the call of cuckoos and nightingales in the trees. And she and Douglas were not entirely alone: Djuna Barnes came to stay, as did the Rickwords and various other friends. Bertrand Russell lived nearby, and Peggy and Douglas went to Petersfield to hear him give a talk in which he warned of the horrors of what he predicted was the coming war. William Gerhardi (author of the novel *Futility*, who garnered comparisons to Chekhov) came to see Peggy, but he and Douglas were at such removes, politically, that he asked to come only when Douglas

---

* The Tolstoys' marriage was notorious for its misery.

was absent. They also saw a good deal of Lorna and Ernest Wishart, both of whom Peggy liked. But she was bored by what the men talked about, which was communism.

Douglas was becoming more and more involved with the Party. It was something of a family concern: Wishart was also a member and Mavin, too, joined during the early 1930s. When Peggy and Douglas argued, now, he called her a Trotskyite rather than a whore. She became irritated by what she perceived as his one-track thinking. "After John's brilliant mind and detachment, all this was too silly for me to endure," she noted. When Douglas gave a course of lectures about the writer as revolutionary, Peggy attended only in order to ask questions afterward that would embarrass and confuse him. He very much wanted her to join the Party, but they would not accept her unless she worked for them. With typical brio, Peggy wrote to Harry Pollitt, the head of the Party, saying that she could not work for them since she was a full-time mother. She could hardly have been refused on such grounds, so she, too, became a member, albeit a most half-hearted one. "All the money I gave him, which formerly went to paying for the building he had done on the house and other things, now went to the Communist Party," she wrote, "I had no objection to that at all. I merely got bored listening to the latest orders from Moscow."

By now the Spanish Civil War had begun. People close to the Garmans were getting involved: Mavin's best friend, Bob Symes (with whom both Lorna and Ruth Garman had enjoyed brief flings), joined the International Brigade against Franco and was killed in action in 1935. Edgell Rickword and Sylvia Townsend

Warner went to Madrid with a delegation of communists, to help with propaganda. Harry Pollitt, too, went to Spain. The Spanish conflict led to rifts within the Garman family. When Roy and Mary Campbell arrived to stay with Lorna, having fled the dangers of Spain, Douglas fell out with them. Edgell Rickword wrote of Campbell, "He was very good fun, by no means a fool. But where he got this crappy, hysterical sort of fascism from, I don't know. At one time I thought his wife might be responsible." By this time the Campbells had become staunch Roman Catholics, and they were on the side of the Right. Roy Campbell called the Wisharts' house Bolshevik Binsted. The Garmans' extreme politics did not make for happy gatherings. Poor mild Mrs. Garman presided over a Christmas lunch which was marred by shouting. "Fascist!" accused Douglas; "Communist!" roared Roy. After this, Mary and her brother were never on cordial terms again.

Douglas bought a second-hand car and drove about the country, lecturing and recruiting. Peggy was left to her own devices, something she never liked. She was more and more alone, and became more and more unhappy. According to her, Douglas was by now so fervent that he became intolerant of anyone who did not adhere to his own creed. The only people he now invited to Yew Tree Cottage were Communists, and it didn't matter what other qualities they had: if they were Communists he welcomed them. "Any person from the working class became a sort of god to Garman," she noted. Worse, he reveled in an asceticism that irked the comfort- and pleasure-loving Peggy. By now, the King had abdicated and so Peggy expected

Douglas to keep his word and marry her. Since they were getting on so badly, largely over his politics, he refused. Peggy was so furious that she went out into the garden and tore up his best flower bed, hurling his rare plants over the fence into the field next door. The next morning, she felt deeply ashamed of herself and enlisted Jack, the gardener, to help her replant them. Jack behaved as though nothing unusual had occurred, loading the plants into a wheelbarrow without comment. Most of the specimens died. "This incident did not help to further my matrimonial aspirations," wrote Peggy.

In fact, Douglas had met someone else: Paddy Ayriss, a Communist remembered by one of Douglas's nieces as the very epitome of a tender comrade; neat, efficient, and with plaits around her head. It is likely that Douglas had already made her acquaintance during his time in Leningrad in 1926. Paddy was there at the same time, having been effectively banished from Communist Party headquarters at King Street in London, where she had had a job as confidential stenographer. In 1925, she began a relationship with a Yorkshireman called George Hardy, who was active in the Party. Hardy was already married and the father of two children. When Paddy discovered in the latter part of 1926 that she was expecting his baby, the Party was keen to get the errant lovers out of the way. There were various reasons for this illiberal attitude. One was that Hardy's wife began turning up at King Street, making threats to expose the matter in the press. Although the Party generally turned a blind eye to the private liaisons of its functionaries, when it came to behavior which risked alienating either party members themselves or the wider circles they sought to influence, they could be puritanical. It was then that George Hardy and Paddy Ayriss were shipped to Russia in haste. In time, Hardy obtained a divorce, and he and Paddy were married.

"Garman was very upset about his new love affair because Paddy had a husband," Peggy later wrote. "They seemed to belong to the Communist Party to the extent of having no say about their private lives. Garman did not want to interfere and break up their marriage, and anyway Paddy wasn't quite ready to leave her husband, who was much older than she was, and whom she rarely saw. Also they had a child. They both belonged to the working class." Paddy was a dedicated worker in the Party cause, capable and energetic. "She was very attractive," Peggy noticed. "She looked rather American, with a tiptilted nose and a smart little figure." Her real name was Jessie Emma, but her childhood tempers had earned her the nickname Paddy, and it stuck. About this time, Douglas, too, was given a nickname. Peggy's children had started calling him Garry (from his surname, presumably), and now Paddy also addressed him thus. Although his siblings continued to call him Douglas, all his political comrades from then on knew him as Garry.

Sometime in 1937, Paddy did leave her husband for Douglas. Peggy kept Yew Tree Cottage, and Douglas's daughter continued to spend weekends there with Peggy and Sindbad and Pegeen. The one thing Peggy and Douglas still had in common was the children. In order to see as much of them as possible, Douglas often stayed at weekends, and he and Peggy generally slept together, making their partings the more painful and difficult. It took them six months to break completely, including an abortive attempt to live together in London. They also spent time in Paris, making love and visiting exhibitions when they were meant to be making practical arrangements for their separation. The cause was never out of his mind, and he was very happy in Paris, Peggy noticed, "because of the *Front Populaire*, which was at its height." Peggy would tease Douglas by saying that Harry Pollitt had sent for her to discuss the Paddy affair, playing on Douglas's concern about disapproval from the leadership of the party. In the end, Douglas moved to Hampstead with Paddy, and they were married.

In her newfound solitude, Peggy needed something to do. She toyed with the idea of starting a publishing imprint, but having decided that this would swallow too much money, she decided instead to open a small gallery selling contemporary art. She began with sculpture: Jean Arp, Constantin Brancusi, Henry Moore, Alexander Calder. Jacob Epstein invited her to look at his work and was cross that she brought Jean Arp with her for the studio visit, for he hated surrealism. For her part, Peggy liked only Epstein's "miraculous" portraits. "They are certainly as good as anything done in the Italian Renaissance; but I hate his other concoctions," she said. In her new London gallery, she showed work by Yves Tanguy, Jean Cocteau, and Wassily Kandinsky, work that went on to be part of the legendary Peggy Guggenheim collection in Venice.

Peggy cured her heartbreak over Douglas by becoming infatuated with a young man she met in Paris through their mutual friend James Joyce. He was "a tall, lanky Irishman of about thirty with enormous green eyes that never looked at you" and his name was Samuel Beckett. She brought him to stay with her at Yew Tree Cottage, where he met Deborah and Lorna, but the affair did not last. In the late 1930s, she left the house for the last time. She later married the surrealist painter Max Ernst, but this marriage also failed. When the conductor Thomas Schippers asked her how many husbands she had had, she is alleged to have replied: "D'you mean my own, or other people's?"

Douglas's marriage to Paddy was an affectionate union, as well as a politically compatible one. In letters from Communist Education Schools in the late 1940s, he calls her "sweetheart," "honey," "darling," "my little honey-bee," and "my little noodle." He certainly depended on her practical nature and reliability. After a decade together, he rued their temporary separation, while celebrating their shared values. "Don't be worried that my work means I am often away from you. We are both working for the same thing, and are therefore much closer together than many thousands of people who are physically never separated. And the times we are together are closer and richer in love as we grow older in love." However, among Douglas's notebooks there is part of a story written many years later, in which the hero meets his former lover's daughter, now grown up, on a train in Europe, a meeting which leads him to reflect that her mother was the true love of his life. And the woman's description matches not Paddy, but Peggy.

Chapter Nine

You Beautiful Creature

The Garmans changed people's lives. According to Mavin's son Sebastian, "most people weren't used to having their reality enhanced like that. They created a magic that could sometimes be destructive." This was true of none more than Lorna. She inspired great love and many imitators, but she also broke hearts. She had a commanding presence and always seemed to be the center of attention, never leaving or entering a room unnoticed, from her early childhood till the end of her days. And she was the loveliest of all the sisters, spellbinding: "The great thing about Lorna is that she had these amazing eyes," recalls Pauline Tennant, whose father, David, owned the Gargoyle Club. "They were profoundly blue. When she looked at you, you felt transfixed. She was *remarkably* beautiful." She became something of a role model to her younger nieces. "Lorna was just glamourous, like a Hollywood star, like Paulette Goddard or someone," Helen's daughter Kathy remembers. "She'd twiddle her hair round and round her fingers while she looked at you and we all copied her." To her daughter Yasmin, she was an irrepressible spirit. "She'd come back from London and lean over our cots, with diamonds and Chanel No. 5, and it was mind-blowing. She was this apparition, even to her children. She was like a force of nature."

A studio portrait of Lorna, circa 1930.

Lorna set out to create magic. She gathered glow-worms from the side of a stream and put them in wine glasses lined with leaves to make natural lanterns, which she'd place all along the mantelpiece. She loved spontaneity and surprises. She went riding on her horse at night, through the steep streets of Arundel, where people were sleeping, a small tame goat following behind. Years later, when she had grandchildren, she would go alone into the woods and decorate a Christmas tree, complete with candles, before leading the children out to find it glowing mysteriously. In the middle of these same woods was a clearing with a pond, and this, too, she transformed into an enchanted glade, with lanterns and her own carvings draped in beads. She painted idealized landscapes, dripping with moonlight.

She loved swimming and would do so anywhere, at any time of year, and long into her old age. She'd strip to her knickers and plunge into thirty-foot waves in Cornwall in the winter or into remote lakes or fast-flowing, icy rivers. She'd swim naked in the shallow, weedy water by Arundel, mocking anyone too prim to take off their things and join her. She was feckless, too, in her generosity, giving away money or jewelry on a whim. She never felt guilty, never felt ashamed. "She was amoral, really," says her daughter, "but everyone forgave her because she was such a life-giver." Being unconventional was almost a point of honor. Lorna drank Guinness at the hairdresser's. She hated crowds. She loved cricket. She was romantic and passionate, but she also had a flinty streak, a heartlessness. She was not without vanity, and she could be cruel, yet she was also capable of great kindness.

Because Ernest Wishart had married her when she was only sixteen, he seemed to accept that she would have love affairs. But their marriage was no sham, and they were devoted to each other. Every day they took long walks together in their ancient Sussex woodland, talking for hours. In later years, when Lorna had converted to Roman Catholicism (a decision influenced by her sister Mary's ardent faith), they would argue terribly, Wishart for socialism, Lorna for religion. Yet there was a wryness underlying their discord, as Pauline Tennant remembers: "Wish was like a russet apple, with red cheeks, benign and silent at the breakfast table in the kitchen. And Lorna came in—she always made an entrance—saying, 'Oh, Wish: *could* we afford a taxi to send Mother Superior to Reading? She's such an angel!' And Wish just said, quietly, 'I didn't think angels needed to ride in taxis.'"

In her early childhood, Lorna had been extremely close to Ruth and Mavin, her two nearest siblings, who had spent their adolescence running wild in Herefordshire together. Mavin adored her and named his only daughter after her. They shared common literary influences: D. H. Lawrence, Wilfrid Scawen Blunt, and John Cowper Powys. Freedom to do as they pleased and to bestow affection where they wanted was essential to the Garmans. What mattered was the freedom to live life to the fullest.

Llewelyn Powys, one of the younger brothers of John Cowper Powys, published *Life and Times of Anthony à Wood* with Wishart & Company in 1932. Like his brothers, he did not confine his literary endeavor to one form, and his essays, memoirs, biography, and fiction reflect his pantheistic relish for life and poetry. He had traveled in Africa and America, and is one of only two writers (the other was his brother John) to have known both Thomas Hardy and F. Scott Fitzgerald. Edna St. Vincent Millay was among his dearest friends. He was an enthusiastic advocate of free love.

To celebrate the publication of his book, he invited Ernest and Lorna to stay with him and his American wife, the writer and editor Alyse Gregory. He advised them to bring very little luggage as the Powyses lived in a cottage accessible only on foot, high on the downs near Dorchester in Dorset. It was a romantic setting, remote and untamed, just the sort of place Lorna loved, and her time there proved to be the beginning of a friendship that was to last for the rest of Powys's life. "It was a most fortunate week-end," he wrote to Wishart, adding about Lorna's characteristic laughter, "I shall never forget that laugh—so free, so lovely, so scandalous that rang so gaily through our home."

Llewelyn liked Ernest well enough, although he would subsequently rue that his publishing company turned down books both by himself and his brother John. But it was Lorna he was really taken with. Somehow the two subsequently found the time and opportunity to become close. Each recognized a kindred spirit in the other. Powys was one of eleven, the children of a clergyman, and like Lorna, he was among the youngest in what was essentially a Victorian family. Both were to react against their upbringing in their advocacy of sexual and personal liberation. Llewelyn, twenty-seven years her senior, was probably the greatest single influence of Lorna's life. They shared an almost ecstatic feeling for the natural world and a relish for the poetic and symbolic: knights, maidens, unicorns, and bacchanalian scenes occur throughout Llewelyn's letters to Lorna and were to become recurrent themes in her paintings. They had in common an unabashed, even bawdy, sense of the physical. They also shared the passionate belief that pleasures should be taken wherever they were offered. Writing to Lorna, Llewelyn quotes a Chinese sage: "Let the eyes see what they want to see. Let the nose smell what it wishes to smell. Let the ears hear what they wish to hear. Let the body enjoy the appetites it wishes to enjoy and let the mind engage itself in every thought that gives it pleasure." These were the tenets which Lorna was to live by.

"All of the Powyses have charm, various charm, but Llewelyn abounds in it incomparably," wrote Louis Marlow in *Welsh Ambassadors*, his biographical study of the family. "His smile alone, with its broad sudden light, is enough to win the stoniest heart."

Alyse Gregory noted, a touch ruefully, that all women fell in love with her husband because of his good looks and distinction, the innocence of his character, his poetic view of life, the delicacy of his health, his essential insecurity, and his fearless intelligence. This propensity to fall in love was reciprocal. Llewelyn was not a faithful husband, nor did the utterly devoted Alyse expect him to be. The doorstep of his house in Dorset was inscribed: "On earth is wine, bread and bed." In Llewelyn, then, Lorna had met her match. He wasn't afraid of her, as her younger lovers were later to be, nor was he possessive. Perhaps uniquely in her life, he did not hold back from offering her advice. His letters to her, written from 1932 until 1939, are the only ones Lorna received which survive, since she burned all those she later had from other lovers.

"Do not be silly but live with intensity," he urged her. "Improve your mind and your manners but let your novels go— Walk by yourself in the winter twilight—Walk at night and in the small hours listen to the winds wild whispering." Powys had weak lungs as a result of tuberculosis, and his illness made him scornful of wasted time. "I hope you gave attention to my scoldings and are much more careful to number your days," he says in his next letter. "I hope your heart only now beats quickly for love." Love was of the greatest interest to him. Imagining Lorna in the transports of erotic delight—whether with himself or with her husband—was a leitmotif of his letters to her. A nephew remembers Lorna laughingly reading out loud a passage in which he reveled at the idea of her, pinioned beneath her husband, and indeed there are several such flights of fancy in his letters. "I hope your good man may treat you to a fanfare making you laugh and cry until you lie on your beautiful shoulders exhausted beyond measure and ready to dream. But let him go about it with prolonged deliberation...I would have you smile all full of love." Another letter says that he "would like to be a sparrow to look in upon the enchanted bed when you lay at last laughing and crying utterly exhausted with his sweet usage." Writing again, he says: "How lovely to think of you lying all languid, ah languid and tremulous on a tanned HAY COCK—I was charmed by the thought of it." Congratulating her on the birth of her daughter, her third child, in March 1939, he writes, "I envy you having these little children. Multiplying yourself by your lovely midnight and midday fooling with your Lord—that fooling that made you little Lorna— egg-white and wild and dark...as a debauched child angel spread-eagled on Caesar's bed...the mind more in flames than the body, sweeter than the sweetest Bacchanalian Bank."

This Chaucerian vein is only one part of their several years of correspondence. His affection and concern for her are just as vivid. He calls her Lady Lorna, ending his letters, "you beautiful, beautiful creature" (the same words Peggy Guggenheim had used to describe her) or "you lovely, lovely Lorna." Wishart was referred to by his initials, "E.E.," or sometimes, archly, as "your Lord." "I pray you always to let me know if you are troubled for if I could lift your burden by the weight of a grasshopper's foot it would be my delight and pride," Llewelyn writes.

It made me sad to know that you were passing through a difficult part of the Forest. You must have confidence and build up your own inner life—trusting to your destiny. You must not be silly and behave foolishly like any other "pretty" girl—it is most unworthy of a proud and well-descended girl... You must build up your life on strong foundations. It is only our own folly that makes us quarrel with it—if we have no physical pain it is to be valued—and how much more for a beautiful girl such as you—Value your days—grow more and more conscious, cherish your own proud inner life and do these things in the simple background of gardening looking after your children...

Sound though this advice was, Lorna was not ready to take it. If her marriage was not exclusive, neither was her allegiance to Llewelyn (and nor, in fact, was his to her. He had at least two serious love affairs after he met Lorna). She saw very little of Llewelyn, and their lovemaking was of necessity seized in the brief intervals when they could be alone. She was still hungry for experience, for love, for more. "I have in my life often observed that lovers possess a strange power of persuading accidental meetings to take place," wrote Llewelyn in his autobiographical *Love and Death.* Lorna possessed just such a power.

During a holiday on the Cornish coast with her children, she met the young poet who was to become her next lover. It was August 1937. Her two sons, Michael and Luke, were prep school boys and she was twenty-six. Her glamour was overwhelming: she smoked and wore strings of pearls with her bathing dress, and she drove a Bentley, fast. Unlike Kathleen and Mary, she had wonderful legs, but like each of them there was something feline about her. The poet noticed that she alternated "between a downcast eye-averted shyness with extreme teasing taunting familiarity."

Her laugh was singular and suggestive, described by Llewelyn as the richest and most wanton laugh that he had ever heard. Lorna's voice has been described as gin-soaked, with a rasp, but it was softer than that, less redolent of the night. Her voice was light but not high, with a strong hint of the Edwardian: she didn't quite say "heppy" for "happy," but such old-fashioned pronunciation wasn't very far off. There was a warmth to her voice, an intimacy. One friend described it as being like crushed velvet.

Even the most casual and unposed family snapshots from these years make her look like a film star: Lorna demure in a Pierrot dress with a big white collar and a ribbon in her hair, or in a pale wool coat, looking wistfully into a pond. Other photographs show Lorna rustic in ankle socks and brogues and expensive tweed with kick pleats, or radiant, sylphlike, and secret with little dark nipples like berries, naked in a Cornish cave. The smell of her was thrilling, too. The single flower perfumes of Victorian days, simple rose or lavender waters, had lately been replaced by a new kind of perfume, in which several different scents were blended.* Chanel No. 5 was a favorite of

---

* Chanel No. 5 was launched in 1921, heralding the golden age of scent manufacture. Blending became an art form: Caron, Patou, Guerlain, and Lanvin were its masters. The base notes were the earthy, almost leathery sandalwood, cedar, and amber, overlaid with penetrating flowers. Lorna also wore Caron's Incre (no longer available), and Kathleen used Patou's Adieu Sagesse (also obsolete), probably because its name amused her. Helen favored Chanel No. 5, as well as a scent called Soir de Paris which came in a blue bottle in the shape of the Eiffel Tower. Kathleen's favorite was Arpège, a sophisticated scent with a slightly dusty, powder-compact, leather and fur and banknote aroma lurking beneath a haze of lily of the valley. Lorna's favorite, Fleurs de Rocaille, is still made, and it smells almost *too* sensual to modern, citrus-accustomed noses—like an unaired honeymoon bedroom. Arpège has been relaunched, but the new scent bears little relation to the original.

Lorna's, as was Caron's heady Fleurs de Rocaille. Compared to the chasteness of floral toilet waters, these new perfumes were almost shocking, with their sweaty animal tang only barely disguised by rich jasmine, ylang-ylang, mimosa, gardenia, and violet. The smell was provocative, sexy, not meant for wives or mothers, but for mistresses. Lorna made great use of them.

Always up early in the mornings, Lorna was out walking alone on the beach at Gunwalloe in Cornwall when she saw a young man playing the fiddle on the strand. She stopped and asked him to play for her. Laurie Lee, then twenty-three, was fair, with a long, sensitive face framed by a fall of straight hair. He had a slender frame and an expressive mouth. He played the violin, wrote poetry, and loved the countryside, having come from the Cotswolds. He was solitary and slightly secretive.

Oddly, Lorna was not the first of the Garman sisters whom Laurie Lee had encountered, by chance, with his violin. On his travels in Spain, two years earlier, he had found himself passing around a hat for coins, after playing Schubert in Toledo's central square. Taking pity on him, for he was badly sunburned, a man at one table asked him to sit down. Lee had already remarked that the little group to which this man belonged was, "curiously striking...immediately noticeable in the ponderous summer twilight. There were four of them: a woman in dazzling white, a tall man wearing a broad black hat, a jaunty young girl with a rose in her hair, followed by a pretty lacy child... The man sat at the table with a distinguished stoop, while his companions arranged themselves gracefully beside him, spreading

Laurie Lee and Mary and Roy Campbell in Toledo, 1935.

their shawls on the chairs and beaming round the darkening square as though in a box at the opera." The man's wife inquired in French if he was German, to which he replied in Spanish that he was English. " 'Ah,' she smiled. 'And so am I.' " After hearing how he had come to be in Spain, they invited him back to their house for dinner, where he ended up staying for a week. His hosts turned out to be Roy and Mary Campbell, who had left Martigues and were then living in Toledo.

Lee was struck by Mary, cool and pale in her dress, with her violet-eyed Celtic beauty. Fueled by wine, Roy was soon boasting about the sensual delights of their marraige, a comic scene which Lee later included in *As I Walked Out One Midsummer Morning*: " 'We never grew tired of it, did we, girl? We must have broken half the beds in town.' 'Roy—please,' his wife murmured, touching her lavender lips. 'I'm sure he doesn't want to hear all that.' " The next day, Lee sat with the poet's wife through the hushed afternoon, watching her finger her beads in the airless shade. "Snatches of England—flowered china and silver teaspoons—haunted the heavy Spanish air."

Spain was to exert a pull on Laurie Lee throughout his life, and in the early weeks of their affair, he told Lorna that he meant to go back to fight on the side of the Republicans as soon as he could. His journey is described in *A Moment of War*, the sequel to *As I Walked Out One Midsummer Morning*. Lorna's powerful allure hangs over the account; never named, she is referred to only as "the girl." This is how he describes the scene: "I told her my plans one evening as she sat twisting her hair with her fingers and gazing into my eyes with her long cat look. She wasn't impressed. Others may need a war, she said; but you don't, you've got one here. She bared her beautiful small teeth and unsheathed her claws. Heroics like mine didn't mean a thing. If I wished to command her admiration by sacrificing myself to a cause she herself was ready to provide one." He tried to persuade her that he would be doing it for her, but this wasn't true, and she knew it. All the same, it was partly their entanglement that drove him on, the feeling of overindulgence and satiety brought on by too much easy and unearned pleasure. He felt guilty, too, that she was married and had two young children. He described her as "rich and demandingly beautiful, extravagantly generous with her emotions but fanatically jealous, and one who gave more than she got in love. For several days and nights our arguments swung back and forth, interspersed with desperate embraces, ending with threats of blackmail and bitter tears, with cries of 'Go, and you'll see me no more.' "

But he was determined. Having crossed to France in September, he got as far as Perpignan before Lorna arrived for a week of passionate farewell. She took him to stay with Helen, who was still living at Martigues. There the lovers spent what he was to describe as "a week of hysteria, too—embracing in ruined huts, on the salt-grass at the edge of the sea, gazing out at the windswept ocean while gigantic thunderstorms wheeled slowly round the distant mountains." He was to remember these days as "drugged with coffee, sunlight, tobacco, pine resin, fever and l'amour..." Later, when he was in Spain, Lorna wrote him "letters recalling the wildly passionate celebration of our last week together, a rapturous, explicit and tormenting farewell."

In December 1937, Lee crossed into Spain. He was imprisoned by the Republicans, who wrongly took him for a spy, before joining the International Brigades in Albacete. From there he went to be trained at Tarazona. In London, a worried Lorna asked Harry Pollitt (the head of the British Communist Party), who was about to set out for Barcelona, to try and find Lee; the two men met in the now disused church at Tarazona, where Pollitt had come to address the troops as comrades. Jacob Epstein was meant to be leaving for Spain, but was refused a visa, thwarting Lorna from giving him a letter, which she supposed would reach Lee more quickly from within Spain. Instead, she sent him pound notes by regular mail, "swoony" with the scent of Chanel No. 5, with which she had soaked them.

She wrote letter after letter in her voluptuous handwriting. In England, rumors circulated that Lorna had left Wishart, and indeed she did stay alone at a flat in Hampstead while she waited for her lover to return. In February 1938, Lee came safely back to London, where Lorna was waiting on the cold station platform.

I arrived at Victoria station and saw the cloud of her breath where she waited. She looked at my hands, then at my face, and gave her short jackal laugh. I sniffed the cold misty fur of her hair.

As we drove north she watched me as much as she watched the road. "Well, I hope you're pleased with yourself," she said. "Didn't give me a thought, did you? I've been through absolute hell—you know that? I even went to a call box one night, a public phone box—can you imagine?—and I got right through to Socorro Rojo, Albacete. Just think—across France, and those frontiers—and all Spain and that war... It took me three hours, and I was crying all the time. I just wanted to talk to you, *talk* to you, can you understand? A man was watching me from a car, and kept giving me money for the phone. No wonder you look so smug."

The final sentences of *A Moment of War* have an intoxicating sensuality. Lee and the girl arrive in Hampstead, where, "She drew me in with her blue steady gaze. I remember the flowers on the piano, the white sheets on her bed, her deep mouth, and love without honour. Without honour, but at least with salvation."

Chapter Ten

THE POET AND THE PAINTER

Lorna and Laurie Lee spent much of 1938 together. Lorna rented a flat in Bloomsbury, at 35 Mecklenburgh Square, which was to be all but demolished by a wartime bomb. Her two little boys were installed with a nanny in another residence in St. John's Wood. Having had two children, Lorna must have been aware of the signs when in November she decided to go back to Wishart. She was already four months pregnant. Perhaps it was to allow her better to negotiate this difficult homecoming that Lee now left the country. Perhaps Lorna hoped, at this stage, to pass the child off as her husband's, although such dissembling would have been uncharacteristic of her. Lee, in any event, appears to have known about the baby. On his way to Cyprus he thought of lighting candles "for the three of us" in a small church on a hill above Athens. He stayed away until the following February, returning to England in time for the birth of his daughter, Yasmin, in March. After Lorna had telephoned the news of the birth to him, he came to the St. John's Wood nursing home where the infant had been born and played his violin among the flowering almond trees outside the window.

For the first few weeks of their daughter's life, the lovers lived together again, in a flat above a greengrocer's in Bloomsbury. But Lorna was finding it difficult to manage, and when Wishart offered to bring up Yasmin as his own, she went back to Sussex. As an old friend of Lee's told him, "in the world as it is, only the rich can arrange to have illegitimate children without inconvenience." Wishart was rich and Lee was not, but it wasn't only about money: Lorna's husband was a father with an established household but her lover was an unsettled youth. Whether Lorna confided in Llewelyn Powys or not, his fictionalized autobiography, *Love and Death*, contains a love story remarkably similar to Lorna's. Dittany, the romantic heroine, "belonged to an age in which passion and grace were attributes more highly to be prized than chastity." Like Lorna, she has a "scarlet mouth, that was curled for singing, and the proud bright head and the careless beautiful eyes."* The hero makes his lover pregnant, and the lover's husband plans to raise the child as his own. Describing the husband, Powys writes, "He had an erratic genius peculiar to his own odd character, his attitude to Dittany's pregnancy was exactly in his style—highly civilised and contemptuous as always of the usual codes." The description fits Wishart perfectly. As good as his word, he was to be a devoted father to Yasmin.

* Powys scholars believe that Dittany is based on Llewelyn's final love, Gamel Woolsey, who was married to the writer Gerald Brenan.

Lorna continued to correspond with her old friend and advisor. Llewelyn's health was failing.

Alyse comes to read to me *Tristram Shandy* at 8 o'clock and we drink tea out of doll's house cups. This is now my happiest moment ... I love to hear the gulls come in from the sea across the dim sky ... The riot of life always entranced me—I never weary of it. I love its savagery and tenderness. It's lust and it's grace. I love to think of the abandoned, insatiable and shameless desires of beautiful women and the refreshment that they give ... Eh! but I hope I shall win back my health again ... Be Happy, Be Happy, Be Happy and it is a state that is most easily to be found in simple places and doing simple things.

By 1936, Powys was forced to leave Dorset for the clearer air of Switzerland, where he had been treated for his lungs years before.

My old doctor seems to be of the opinion that I will recover again as I have the leap of a salmon out of the torrent of dark death ... It is strange to look out at the mountains again—the same as I looked at when I had not a single grey hair on my head. Now listen Lorna you beautiful creature—Read and cultivate your mind and learn taste (so that you will never again be contemptuous of Titian's *Dionysos and Ariadne*) as well as indulge your lovely body—and build your life on the firm ground of simple ordinary pleasures, your children, your garden, the delight of solitary nocturnal walks ... for your own joy and not to pleasure the world.

After Yasmin's birth, Llewelyn wrote to congratulate Lorna, adding a hint that he was not unaware of the circumstances of the baby's arrival. "Alyse sends her love we often talk of you both. I am at last getting better If I am not tumbled in the snow by the Germans and if I have patience I have hope of recovery for a blackberry summer. And I hope I shall see you again and hear your laugh. I love your free nature and you are beautiful. I did think it was lovely of you to read *Love and Death*—I shall never write a better book—never, never—how could I? and I love to think of *you* reading it—every word of it..." Along the side of the page, he added: "Darling Lorna I do not forget you ever." These were his last words to her. In November he died. A much-thumbed picture of Llewelyn was stuck into one of Lorna's photograph albums, alongside snapshots of her family; the only image of a friend or lover she appears to have kept until the end of her life.

Going back to her husband had the contrary effect of enabling Lorna to maintain the affair with Lee. Now that a nanny was ensconced in Sussex, Lorna was free to visit Lee in London, unencumbered. There was a new, giddy secrecy about their meetings, presumably because Wishart had made it a condition of taking her back that she stop seeing her lover. Lee would meet her from the train at Victoria: "Tall in her dark-blue slacks and jacket, brown-cheeked, bright-eyed, giggling like a schoolgirl as we walk arm in arm to look for a taxi." Sometimes he would come back from work with the GPO (General Post Office) film unit to find her waiting for him. "Again everything was forgotten, she hung her head or twisted her hair, and talking or embracing, there was little to show that I had ever flown to Greece, or Yasmin had been born."

Inspired by Lorna, Lee wrote poem after poem. Lee also admired her artless drawings and envied "*her* power of writing which she doesn't seem to realise is always warm and brilliantly original." She would arrive bearing gifts: lilies and orchids, "some little white flowers, their petals beautifully marked with lipstick," long, slender red roses, gramophone records of Brahms, Chopin ballades, Debussy, Beethoven quartets. She brought him magnolias, Chinese poems, lilac, books of Stendhal and John Donne. Her every movement captivated him. "L puts on her nightdress with such a soft slow twist to her body that I think of salmon and the play of rivers and the gentle gestures of smoke. I sit hunched in bed watching her silent grace, a woman, alive, adoring the hour, the closed door, my eyes and the promise of sleep," he wrote in an unpublished diary of the period. Lit by a gas fire, they would lie together in Laurie's rented room in Putney, oblivious to the blackout that darkened the streets of wartime London. "I cannot think why lovers ever leave their beds," noted Laurie.

In fact, the two often left theirs for the woods and fields of Sussex and the green hillsides of Laurie's native Gloucestershire: "Everywhere blossoms, orchids, moon daisies, roses, the garden a heady pool of syringa, the banks and fields vital with scent and sound. Nightingales in the woods. I lay with my face in the grass, or with my mouth to hers. Everything was the beautiful familiarity of home, she and the valley were never divided." Once, while they were in the woods, enemy planes came under fire directly over their heads. "With our faces together we waited for the next, our lips were warm against each other, the machine-gun cracked again and with our lips still touching we stared into the sky in each other's pupils, incredulous but without any surprise." On subsequent visits to the Cotswolds, they would go to Stroud, have lunch, and return to lie naked by the lake or in the woods or the fields. Visiting Chichester for a concert, they lay out in the sun by the sea. "The act of kissing in public seemed to

Lorna with Yasmin, winter 1941.

increase our pleasure. I know she is mine by the smell of her mouth, the shape of her arms, and that intangible, free flow of her soft body when she embraces me." With no home to go to, they often ate on their feet: beechnuts on the downs, fish and chips while walking in the Cotswolds, peanuts in the street "till we rattled." Sometimes they would cycle to the Chanctonbury Ring and lie together there on the coarse springy grass. "We stood against the sky touching each other and she looked into my face, laughing and saying, 'You are extraordinary, you are really.'"

By the end of October 1940, Lee had decided he would get out of London and find a shack to live in near Lorna in Sussex. By luck he found just the thing, a caravan (not, like Augustus John's, a picturesque Gypsy caravan, but a simple green tin lozenge) lit by oil lamp, which filled the small space with its sweltery smell. Lorna evidently relished this D. H. Lawrence-like arrangement, arriving in her dramatic Bentley with her arms full of presents, cooking aromatic Mediterranean dishes in the little kitchen area, lying across the bed in the flickering light.

Lee noticed everything about her. He admired the delicate outline of her skull—"I cannot imagine it ever getting old"—and the distraction in her eyes when she thought of her children. She would wear a green tartan frock "with a long line of se-ductible buttons down the front," a black sweater which matched her black hair, a summer dress "scarlet, like a pimento," a green dress, a little white scarf, jodhpurs which had a "peculiar effect" on him. There were furs and a sealskin coat for which, she could not resist telling him, Wishart had paid fifty pounds.

There was a camel cloak; a scarlet corduroy suit; an exquisite wine-colored coat, which made her glow; a wide-brimmed black straw hat "in the shade of which she's a dusky beauty." Once she turned up "in seven different blues (counting her eyes)."

To the caravan Lorna brought books, eggs, tea, sandalwood, and honey. She was endlessly generous, arriving "like a stroke of good fortune, showering presents." She brought flowers, mushrooms, and bananas;* a tambourine, shells, and a chessman she had carved herself from apple wood; Beethoven's *Spring* sonata; rabbit, and vegetables; a bag of sweet cakes and two bottles of wine; snowdrops, Christmas roses, and Champagne; eight goose eggs "and an eddying fragrance of irresistible passion." She came bringing an Oxford book of verse, her hair becomingly piled up on top of her head, fixed with a large comb she said was her grandmother's. "Her face was fine and lovely under the sleek ebony of her coiffure; when she had posed and held her head gracefully for me [he sometimes did drawings of her or took her photograph], she became impatient, combing it all out and saying I couldn't get my fingers into it. Which was true, but I did like it. We played some old records (Satie, which overwhelms us with its tenderness). The storms and the tortured trees were forgotten, the afternoon became one with the night." She left love messages for him, hidden inside the china cupboard.

Eventually, the caravan felt too small and remote, and Lee took a cottage in Bognor Regis. He and Lorna were tortured by

---

* Bananas were a rare treat during the rationing of the war and afterward, so much so that Evelyn Waugh has entered history as a rotten father, for eating two whole bananas, without sharing, while his children looked on.

the furtiveness of their affair and by not being often enough to-gether: "When can I take my girl into the open? ... I have never been able to take her and show her to a friend and say look, this is my woman. I have had none of those normal domestic gaieties of love. Always we have lived in a state of subterfuge, each sec-ond together stolen from the cannon's mouth and consumed under a canopy of uneasiness. For four years it has been like this, for four years I've dreamed about her husband." Sometimes they fantasized about getting married, but neither of them took the idea very seriously. "For five days I'd get two meals a day then for five months I'd have to cook myself," he wrote. (In fact, Lorna did cook for her family all her life, and very well.)

Jealousy plagued them both. Lee was tormented by the thought of Lorna making love with Wishart, and she sometimes taunted him by praising her husband, saying how marvelous he was in a crisis, so dependable, so generous. "She has what she wants when she wants it and everyone else modifies their desires to fit in with her," grumbled Lee, in an apt description of her na-ture. Despite being the one who was married—or perhaps be-cause of this firsthand knowledge of daily deception—Lorna's jealousy was even more intense than Laurie's. She read his diary when he wasn't looking, imagined he was secretly seeing "revolt-ing blondes," made terrible scenes and accusations. She wanted him all for herself. "I'm longing for you to grow old," she told him, "each new wrinkle on your face is a joy to me—I want you to grow old and repulsive so that no one will want to look at you."

They were both aware, too, that the adversity of their cir-cumstances fueled their love. "The minute you step on the train you become a thing of extraordinary glamour," Lorna told him. Privately, he concurred. "It is a paradox," noted Lee, "that the best way to preserve the thrill and intensity of love is to put as much distance as possible between myself and the beloved." Lorna asked her husband if Laurie could come and live with them, to which he replied: "Don't be preposterous. Won't you ever grow up?" Greater distance presented itself in 1942, when Lee went back to live in London, to take up a job at the Crown Film Unit. This was a blow to Lorna. "It's like what I imagined death must be," she said, "coming without warning, cutting short our lives." Lee was making a name for himself as a poet now, having published verse in *Horizon* and *Penguin New Writing*. He had broadcast on the BBC, met Dylan Thomas (who "looked about him joylessly with wet subterranean eyes as if the whole world was a wet Monday"), and become friends with C. Day-Lewis and Rosamond Lehmann. He was enjoying the society of his peers.

Lorna visited him in London, where they walked and went to the cinema. They saw *Casablanca*. Listening to Mozart's flute music in bed one morning, Lorna wept, staining the pillow black and red with eye kohl and lipstick. The affair was as consuming as ever, yet when, in the spring of 1943, Lorna announced that she was leaving Sussex, he found himself dreading her arrival. "She says she'll move to London: In fact I would definitely rather she didn't come. And the prospect of seeing her every night! Isn't that queer after all." Nevertheless, he met her at the station, finding her "brown, nervous, with bunches of flowers," and accompanied her to the room she'd rented in Bute

Street, a stone's throw from the Natural History Museum. Like her sisters, Lorna was able to transform rented rooms. "She instantly made it charming," wrote Laurie, "filling it with vases of flowers, Saint Teresa, a crucifix, pictures of Yasmin, pots of face cream, Rilke, Shakespeare, and the Bible full of pressed flowers." She took a part-time job, at a children's nursery.

It was almost six years since their fateful meeting at Gunwalloe. "I sometimes think with real pride that no one shall touch me but you till I die," Lorna had told Lee. But things changed once she had the rented room. There was a new brittleness about her, a glimpse of something hard and coarse and dangerous. Over dinner in an expensive restuarant, Laurie saw her in a new light, as someone with a hard heart, who drank gin and talked about gigolos and Bentleys, "a world as unreal as hell." He felt "a sort of arrested disgust." On more than one occasion, he went at night to Bute Street to find the room was empty: "Honeysuckle which she had brought from Sussex was thick in the air. I lay on the bed and slept. She did not come in till dawn." Someone else had touched her, and that someone was a young painter, Lucian Freud.

Lorna first met Freud in the summer of 1942. The Wisharts were by then in the habit of taking a wooden beach house at Southwold, and the Freuds also had a summerhouse nearby.*

---

* Lucian Freud's daughter, the writer Esther Freud, made this summerhouse the setting of her novel *The Sea House*.

Lucian Freud had been born in Berlin in the winter of 1922. His father, an architect, was the youngest son of Sigmund Freud, and he moved his family to Britain in 1933. Freud had attended the liberal and artistic Dartington school (Peggy Jean Epstein was there at the same time), then went on to the Central School of Art for a term before leaving London to study at the East Anglian School of Drawing and Painting, in Dedham, run by Cedric Morris. The school building burned down in 1939, probably because Freud and a fellow student had been careless with their cigarette butts. In any event, he took a break from art school to serve on the North Atlantic convoys for a few months during 1941, before resuming his studies at Morris's school, which had relocated to Benton End, near Hadleigh in Suffolk.

A handsome young painter named David Carr was a fellow student of Freud's at Benton End and had been the other smoking culprit in the art room at Dedham. He had become a boyfriend of Lorna's, and it was on a visit to Suffolk to see him that she took up with Freud, in 1943. "Lucian was a real star turn," recalls the painter John Craxton, Freud's great friend of those years, "very, very good-looking, witty, amusing, clever, fun to be with. He was neither English nor German; he found England very exotic. He was *déraciné*, he wasn't bound by conventions. He was very free. And so was she. Lorna was the most wonderful company, frightfully amusing and ravishingly good-looking: she could turn you to stone with a look. And she had deep qualities; she was not fluttery, she wasn't facile at all. She had a kind of mystery, a mystical inner quality. Any young man would have wanted her." The word which people who knew Freud at the time use to describe him is the same one that crops up repeatedly about Lorna: magical.

Lucian Freud's first work to appear in public were drawings in an exhibition of children's art at Peggy Guggenheim's gallery in Cork Street in 1938. But the first public glimpse of his adult work was a drawing* that was published in *Horizon* in April 1940. In this, the fourth issue of the magazine, there was also a poem ("A Moment of War") by a new writer, Laurie Lee. That both of these men should have made their debuts in the same issue of the same magazine would not have surprised Lorna, who was such an inspiration to each of them. Indeed she can be credited with the publication of Laurie's poem, as she had written about him to the poet Stephen Spender, who was involved in the first issue of the magazine. Spender had shown the letter to Cyril Connolly, who edited *Horizon*, and both of them had written encouraging letters to the young Lee, agreeing to print his work. Lucian Freud's drawing had come to the magazine through Peter Watson, *Horizon's* highly influential art editor, who was a collector of note, as well as the magazine's chief backer.

In the same year that he began to see Lorna, Freud took a flat in Delamere Terrace, by the canal in Paddington. The area was rough then. "Delamere was extreme and I was conscious of this," he recalled. "There was a sort of anarchic element of no-one

---

* The drawing is labeled "Self-portrait," perhaps by mistake, since it looks nothing like the artist. A drawing from the same period, supposedly of Stephen Spender, is much more like the young Freud.

working for anyone." Some of Freud's neighbors were violent types, but a taste for the dangerous was tangible in Freud and only added to his already considerable allure. The photographer John Deakin called him "such a strange, fox-like person." Laurie Lee referred to him as dark and decayed looking. "This mad unpleasant youth appeals to a sort of craving she has for corruption," he noted. Jacob Epstein was to call him "that spiv." Much of Lorna's time with Laurie Lee had been a kind of rustic Eden, taking the little Yasmin for strolls by the river, walking in the hills, kissing in the woods. With Freud she spent more time in nightclubs. The critic John Richardson remembers seeing them at the Caribbean Club, a fashionable Jamaican bar-*cum*-nightclub at the back of the Regent Palace Hotel. "I used to see them there, smooching together. Always in the dark. She was very beautiful, very melancholy. There was a sort of sadness about her."

One evening, Lee saw them walking along the street, Freud's head inclined toward Lorna's shoulder. Decades of gossip have transformed a hissed encounter at a Piccadilly bus stop into something approaching a duel. Whether blows were exchanged or not, Freud emerged the victor, even though it was Lee with whom Lorna went home that particular night. Lorna later enjoyed recounting how, fearful that the two might hurt each other, she had called out to a passing soldier, "Can you stop them?" and he had replied, "I expect you're the trouble," and just walked on. "The trouble is he's falling for me. It isn't fair of me I know," she said of Freud, lying on the bed while the miserable Lee sat at her side.

There were scenes. Lee ambushed her at Victoria railway station, snatching her handbag, sure she was hiding something in it.

He haunted her room, frequently finding it empty. That August was the sixth anniversary of their first meeting, but by early September their affair was clearly over. "She doesn't know how long it will last," Lee wrote of her infatuation with the young painter. "She would like to be free of it but can't. Meanwhile she says she loves me." With consummate cruelty, Lorna went with Freud to Cornwall, the place that was so special to her and Laurie. She became, as Lee had once described himself, "light-headed, detached and heartless." Lee put a razor blade to his throat but thought better of it. In October, he went to stay with Rosamond Lehmann, whose son-in-law later recalled Rosamond telling him that Laurie would sit all day at the typewriter, bashing out the same word—"Lorna"—over and over again. *The Sun My Monument*, his first collection of poems, published in 1944, was dedicated to Lorna. Laurie Lee's biographer, Valerie Grove, says that he wore Lorna's signet ring until the day he died.

Lorna might have had a new lover, but the practical details of her life were much the same. She spent most of the week with Freud, as she previously had with Lee, returning to Sussex for weekends with her husband and children. Sometimes she brought her eldest child, Michael, to London with her, treating him more as a peer than a son; she had been so young when she became a mother that Michael sometimes felt that they grew up together. She took him to the cinema with Lee, and later, when she was with Freud, Michael occasionally slept on the floor at Delamere Terrace. But taking the children to London was not her habit. Kathleen's daughter Kitty remembers, when she and her cousin Michael were both eleven, Lorna setting off for

London from Binsted, "and Lorna looked at us and said, off-hand, 'You are a strange pair of children, neither of you living with your mothers, your mothers always going off.' And I remember thinking, in that helpless way of a child, Well we can't help it. It's you who go, not us." A friend recalls that Lorna used to neglect Michael terribly and then have her lovers virtually next door to him (echoes of Vita Sackville-West, who slept with Mary Campbell while her son lay in a bed only a few feet away). In later years, Michael, who became a painter of renown, would boast of his mother's conquests, and he was said to have been dismayed when she destroyed her love letters. In his autobiography, *High Diver*, he wrote, "Suddenly, my mother vanished for years . . . I thought it was my fault that she had gone. There followed a time of grief and anguish. Then my mother made brief reappearances. They tore open the wounds of her absence."

It had been convenient for Lorna when, during the affair with Laurie, he had left London and lived near her, first in the Sussex caravan and then at Bognor. So when one of the cottages on her husband's estate became vacant, she thought of installing Freud in it and began to do the place up in readiness. But this didn't come off, probably because her husband put his foot down, and instead she and Freud continued to meet in London. Lorna would bring him things to paint. There was a dead heron she found in the marshes and a stuffed zebra's head from Rowland Ward, the taxidermists in Piccadilly. Freud called it his prize possession, and it appeared in several of his pictures, including *The Painter's Room* of 1943–4, in which the zebra's head, painted in red and cream, emerged through a window into a room containing a battered sofa, a spindly plant, a top hat, and a piece of red cloth. The painting was wet on the easel when a buzz bomb landed close by, shattering the window but leaving the picture undamaged. Lorna bought the picture from Freud's first one-man show at the Lefevre Gallery, in 1944. At fifty pounds, it was the highest price he had earned for a picture. "I think she was Lucian's imagination," says her daughter, Yasmin.

It was around this time that people noticed a change in his work. "It was in 1945 that Lucian Freud began to handle oil-paint in a specifically adult way," observed the critic John Russell. "This was due to a new density of involvement with individual human

*Lucian Freud with the zebra head
that was a gift from Lorna.*

beings ... he had led until then a life un-anchored to specific attachments." The specific attachment, Lorna, began to appear in his work. He drew her in her ocelot coat, wide-eyed. She is the *Woman with a Tulip* of 1945 and the downcast *Woman with a Daffodil* of the same year. Lawrence Gowing later asked Freud who the model for this picture had been, to which he replied, "She was the first person who meant something to me," without revealing her name. "I was more concerned with the subject," Freud said later, "—she was very wild—than I had been before." He later told the critic William Feaver: "Everything is autobiographical and everything is a portrait."

In the catalogue to his major 2002 retrospective exhibition at the Tate Gallery, almost sixty years later, Freud at last allowed his early model to be named. He himself referred to her as "the first person I got keen on," adding that everyone had liked her, including his mother. Although Lorna has been described as a vamp, her magnetism was wider than that term allows, making her attractive to other women and to children. Laurie Lee's mother, Annie, for instance, was very fond of her, and the two women wrote to each other for years after the affair with Lee ended, with news of little Yasmin.

The end of the affair mirrored its beginning, with tears and faithlessness. Freud began seeing another woman, a ravishing actress who was younger than Lorna. Michael Wishart described the woman in question descending the mirrored stairs at the Gargoyle Club, looking "as beautiful as it is possible to be ... in an irresistible scarlet dress, her blonde tresses flying." One day, when Lorna was in the country, she telephoned Freud, and this new woman in his life answered the call. When she heard the woman's voice, Lorna began to cry. His new girlfriend asked Freud who it was, but he didn't tell her, only saying, "it's nobody," before he put the receiver down. Soon after, Lorna found some letters at Delamere Terrace which confirmed her fears. Distraught, she left him.

Freud was deeply unhappy and pursued her to Sussex, where he went to stay with Peter Watson. On one occasion, he arrived at Binsted with a gun, saying he would shoot Lorna or himself there and then if she would not come back to him. He stood in the cabbage patch and fired the gun, but no one was hurt. On another visit, he jumped on a great white horse in the field beyond the lawn and tilted, bareback and at breakneck speed, toward the windows where she could see him. He besieged her by turning up at the house, forlorn and reckless, bearing gifts. Animals had always been a motif of their affair, and Freud now gave Lorna a white kitten in a brown paper bag (she called it White Puss, and it became the subject of several of her paintings). People who loved her called her Lornie; Freud called her Lornie-bird. He kept birds of prey, first a kestrel and then sparrow hawks, one of which he used to take about London with him on the Underground. As William Feaver has remarked, these birds accorded with Freud's image of the fierce, not to say predatory, man about town. When Lorna refused to come back to him, he dumped the zebra head she had given him on the landing outside his room.

"Lornie had much more common sense than the rest of the Garmans," says Yasmin of her mother's resolve to part from Freud. "When the chips were down, she made good decisions.

Her feet were on the ground." During her time with Laurie Lee, Lorna had begged him, "Make me not want you so much. I can't get any peace always thinking about you—I want to be able to live in peace again." She had been tormented, too, by her feelings for Freud. But after the argument caused by finding the letters, she told him, "I thought I'd given you up for Lent but I'm giving you up for good."

The woman with the blonde tresses who became involved with Freud remembers that the bed at Delamere Terrace was on a high platform and underneath it, on the wall, someone had written words from Shakespeare: lines about being powerless, enslaved to the beloved. Some years later the woman in question was fascinated to meet Lorna, who told her that it had been she who had written the words from Shakespeare on the wall, she who had telephoned in tears that night. The woman asked how she could now bear to talk to her, to which Lorna replied, "Because you saved me from a terrible obsession."

But the men who had loved her found it hard to get over Lorna. More than a year after they had ceased to be lovers, Laurie Lee still fantasized about killing her, and about suicide. A contemporary recalls having drinks with Michael Wishart and Lorna at the Café Royal, a year or so later: "And Lucian came up but wouldn't sit down and join us, and I could see that he was trembling." "Lucian was genuinely in love with her, but she never went back to him," says John Craxton. "It was the great love of his life. He said to me—I've never forgotten—'I'm never, ever going to love a woman more than she loves me' and I don't think he ever did again. He never really forgot her. He wrote letters saying 'I still love Lornie and miss her.' She *was* a muse, a true muse in the best possible way."

In 1946, Lucian Freud spent two months in Paris, then joined Craxton for five months in Greece. In 1948, he married for the first time. His bride was a wide-eyed twenty-one-year-old with heavy dark hair and a beautifully curved mouth. He painted her picture again and again: she is the *Girl with a Kitten* (a pun on her name), and the *Girl in a Dark Jacket*, both from 1947; also the mysterious *Girl with Leaves* and the *Girl with Roses* of 1948. She is the subject of one of Freud's finest pictures, the *Girl with a White Dog* of 1950–51,* now in the collection of the Tate. The English bull terrier nuzzled in the crook of her knee in this picture was one of a pair given to the couple as a wedding present. The girl was Kitty, the daughter of Jacob Epstein and Kathleen Garman.

Laurie Lee married a couple of years later. Since the split with Lorna, he had become a regular visitor at Kathleen's house, 272 King's Road, only a five-minute walk from Markham Square, where he was lodging. Ostensibly, he came to play his violin with the other young musicians who congregated at 272, but it became obvious that it was not only the music that drew him to the house. Mary Campbell's daughter Anna, who was staying with her aunt, was the first of the younger generation who caught his fancy, and then he turned his attention to her cousin Kitty, before her marriage. Helen's daughter, Kathy, also staying at 272 by then, was younger, fairer, and more sparkling

---

* In 2003, a reproduction of this picture became the Tate's best-selling postcard.

than her cousins. She knew that Lee was the man for her. When he took one or other of her cousins out for the evening, she used to put his coat across the end of her bed, so he'd have to come into her room to retrieve it before he went away.

Lee had recovered his spirits by now and was on friendly terms with Lorna. She wrote him letters that were intimate yet breezy, littered with kisses and with news of little Yasmin. At a party given by Roy and Mary Campbell in London, Lorna took him to one side, pointed to Kathy and said, "Try that one. She'll do for you. She'll be strong enough." By the time Kathy was sixteen or seventeen, he began to call for her, taking her to the cinema and writing her poems. In 1950, Kathy had been away in Italy, and when Lee met her at Victoria Station (scene of so many of his meetings and partings with Lorna), he noted that she was "more radiant than ever, but twice the size." He took her straight to the scales, saying he wouldn't marry her if she weighed over twelve stone (168 pounds). She was not, so he did.

Both Lorna's lovers had found a wife, and each bride was one of her own nieces.

Chapter Eleven

LADY EPSTEIN

In the war years before Kitty's marriage, Kathleen stayed in London, despite the blackouts and the real danger of bombing. She remained at 272 King's Road with her sister Helen, who had left Martigues at the beginning of the war, fearful for her daughter's safety. Helen had begged her husband to come with her, but instead Polge stayed on to join the French navy. His daughter Kathy remembered waving good-bye to him through the plane trees. Back in England, the little girl was billeted at Vine Cottage in South Harting with her Garman grandmother, while Helen worked for the Free French in London and lodged with Kathleen. A lady who had a room above the pub in Mrs. Garman's village came to give Kathy French lessons and encouraged her to write to her father. "I am not afraid of the wolves," he told her in a letter. One day in 1942, a telegram came. Helen held it in her hand while she cried: Polge was dead. He had been killed on active service, but his daughter ascribes his early death to another cause. She believes he died from sorrow, that his wife and child were not with him.

At weekends, Kathleen and Helen would take the bus from London to Sussex, where Kathleen's children, too, were lodging with their Garman grandmother. As usual, Kathleen conjured up fun and excitement for everybody. She somehow managed to obtain a hut half a mile from the village, and here the children had their own little gardens, a strip for each child. Kathy remembers growing zinnias and lettuce in hers. There was a piano and they all played music. "In the rainy afternoons and evenings we'd stay on at the hut and read Shakespeare plays, and Chekhov and Ibsen," says Kathy. "My mum would be exhausted from working so hard all week, but my aunt was a great 'Let's do that!' person. She gave us piano lessons and we all learned speeches off by heart."

Kathleen's enthusiasm was not always welcome, however. Kitty dreaded these visits, embarrassed by her mother's "Londony" clothes and alarmed by her: "My mother would set us things to do. She'd set out things for us to paint. There were dried figs in green paper. My mother would say: 'What can you do? What are your talents? Can you play the cello?' And of course I couldn't because no-one had bought me a cello or given me any lessons. She wanted me to excel, but in a way she didn't. I think she wanted us—her daughters—to excel, but she didn't want us to succeed, because she had to be the queen. She wanted Theo to succeed. I was frightened of her because of her temper and she did say searingly sarcastic things." Sometimes she was unaccountably unkind, as when Kitty had her tonsils out in a London hospital: "She and my father came to see me, she in long evening dress and jewellry, on their way to dine in Soho. My mother criticised me for being a cissy and needing a separate room, and tried to persuade the nurses to move me to the larger ward. She succeeded, and when I cried because of my sore throat, the other children jeered."

For a year or so, Kathleen stayed in Harting for an extra day or two each week, teaching the piano at a local school. She is remembered for wearing a leopard-skin coat and matching hat and an enormous, dome-shaped ring on one hand, a sapphire or an amethyst. She was known locally as the Tiger Woman, with deliberate echoes of the racy novelist Elinor Glyn, whose novel *Three Weeks* had caused a sensation with its depiction of love on a tiger skin.* One pupil was absolutely terrified of her. Having previously loved the piano, she now refused to practice, too frightened of Kathleen to go on with her lessons. Fortunately for Kathleen, there were other pupils. Another Harting resident recalled that there was always talk about the Garmans in the village, because of their "funny set-up and the Epstein business. They were not considered quite respectable, in those days." If this mattered to the once utterly respectable Mrs. Garman, she did not show it. Everyone was welcomed into Vine Cottage.

Despite the shortages of wartime, Kathleen still found people to do her bidding. On the bus on the way to Petersfield she met a Frenchman who played in a palm court orchestra in Portsmouth. "My mother made friends with this man, a Mr. St. Angelo," recalls Kitty, "and arranged for him to come to Hart-

---

* Elinor Glyn coined the word "it" for sex appeal and inspired the lines "Would you care to sin / with Elinor Glyn / On a tiger skin? Or would you prefer / to err with her / On some other fur?"

ing to teach Essie [Esther] and me French. She gathered up people wherever she went." In London, too, Kathleen befriended people, especially foreigners or those whom she felt were outsiders. A family living in Duke's Lane, off Kensington Church Street, became close friends. These were the Vivantes, refugees who had been forced to leave Fascist Italy and their beautiful Tuscan house, the Villa Solaia. Leone Vivante was a philosopher and literary scholar, and his wife, Elena, an artist. Elena's brother, Lauro de Bosis, had been a poet and antifascist hero, who had died in 1931. In London, Leone wrote a book, *English Poetry*, which was published by Faber & Faber, while his wife joined the Italian section of the BBC. The Vivante children were to become great friends with various of the Garmans, as were their parents. After the war, the Villa Solaia became a favorite port of call during Garman visits to Italy. It was through the Vivantes that Helen met Mario Sarfati, who was to become her second husband in the late 1940s.

Epstein still called on Kathleen twice a week, on Wednesday and Saturday, often with his arms full of her favorite flowers. They dined out at the Caprice, the White Tower, or at a favorite Italian restaurant of his, Ciccio's, in Kensington Church Street. He stayed the night at 272 before leaving very early in the morning, tiptoeing down the stairs with as much stealth as he could. To supplement their two evenings together, he generally came to spend an hour or so with her after work, late each afternoon, but neither of them was inclined to turn up unannounced on the other's doorstep. Mrs. Epstein seldom left Hyde Park Gate, and Kathleen would not have wanted the indignity of meeting her, with or without the pearl-handled revolver of their earlier encounter. The formality of Kathleen and Epstein's arrangement allowed each of them considerable freedom. It also enabled them to keep secrets from each other.

Visiting an exhibition of Epstein's drawings with Lorna in 1939, Laurie Lee had noted, "Some from *Les Fleurs du Mal* unlovely and obscene. Several fine portraits of a repulsive little boy called Jackie, some lovely nudes of a thin negress, and several variations on love and death." The "repulsive little boy called Jackie"—actually a very nice-looking child—was Epstein's son, born in 1934. His birth mother was Isabel Nicholas, who had arrived at Hyde Park Gate without preamble in the summer of 1932, offering to pose as a model. She was an art student who wanted to become an artist, she had no money, and she was very beautiful. Mrs. Epstein welcomed her into the household, hoping as ever that someone closer to home might usurp Kathleen. Before long, Isabel had abandoned her own bedsit in Holland Park and moved in with the Epsteins. She shared a room with Epstein's daughter Peggy Jean, went to the cottage at Epping Forest with the family, and even joined them on a trip to Paris. One of Epstein's very best pieces, the haughtily sensuous *Isabel* of 1933, depicts her bare-breasted, wearing vast, squiggly earrings. But although she and Epstein became lovers, his heart still belonged to Kathleen. While with Isabel at Epping, he wrote to Kathleen, who was away at Martigues. "Apart from working, being with you sweet girl has been and is the obsession of my life & now you are hundreds of miles away . . . I can see you are surrounded with loving friends but remember I am here and thinking daily and hourly of you . . . I can only regret this parting and until I see you I cannot be happy."

Epstein did not tell Kathleen when Isabel became pregnant, nor when Jackie was born. After the birth, Isabel went to Paris for a time. There she married her first husband, the foreign correspondent Sefton Delmer, and sat for André Derain. Alberto Giacometti became infatuated with her, and Picasso painted her portrait five times. She later married the composer Constant Lambert, designing sets and costumes for his ballet, *Tiresias*. After he died, she married his friend and fellow composer Alan Rawsthorne, and it was as Isabel Rawsthorne that she became the friend and frequent model of Francis Bacon. Daniel Farson wrote that she enjoyed "a life of uninhibited exuberance until the onslaught of old age," wearing "the surprised expression of someone who has just heard the most marvelous joke and wishes to share it." Friends remember her as funny, original, spirited, intelligent. But no one who knew her ever mentioned the child. Neither did she.

Margaret Epstein once again stepped in, taking the baby on and bringing him up as her own, as she had already done with Peggy Jean. She even began to lie about her age, so as to make it biologically possible that the boy could have been hers: when she died, at seventy-four, her death certificate gave her age as sixty. The boy had successfully been kept a secret from Kathleen for the first five years of his life. She only learned of his existence from a piece in a newspaper about an exhibition of Epstein's. She is said to have turned very white, whispering, "I've just had some dreadful news."

Epstein was passionately jealous where Kathleen was concerned, even as he remained married and took at least one other lover. "Epstein was 100 per cent jealous, always. That was something Kathleen had to cope with," a friend recalls. "Once a composer touched her on the shoulder one evening at a party. Literally a touch: that was all. Epstein spotted it from across the room. 'Come on Kitty, we're leaving,' he said. Later, he quizzed me at length about the composer." Kathleen herself took a more pragmatic view of Epstein's fondness for women. Roland Joffe, who was informally adopted by Kathleen in the mid 1950s, remembered a holiday in Scotland with Kathleen and Epstein. "He was in a bad mood until we went to see *Playboy of the Western World* at the Pitlochry Playhouse where there was a ravishingly pretty actress. We went to the Playhouse every night of the holiday for ten days after that. Kathleen was happy that a pretty girl had cheered him up."

In fact, Epstein's intuition that Kathleen had cause to make him jealous was not always unfounded. During the 1950s, a younger Irishman became, in her daughter's words, "rather a boyfriend of my mother's." Before then, in 1944, she began a love affair with a handsome young Italian who was an undergraduate at Cambridge. There were picnics in the water meadows at Cambridge, boating trips, meetings in London. She also saw him, over the coming years, in Italy, where he met her from the train from Paris. Volumes of poetry, in English and Italian, were exchanged as gifts, and he wrote poems to her and drew her portrait. Kathleen burned all her letters when she came to see him one day at his room in college, but he had copied out one, which illustrates the tenderness of her feelings:

What shall I start with, my darling? With today which is soulless and dreary as it should be or with yesterday when all nature sang to us with joy? Or with Cambridge and St Anthony wrestling with his demons or with London and my sad heart lightened up only by the shining garland? Yes! I still walk engarlanded and though of course it is invisible to others, the reflection must be in my eyes for they keep asking why they are so bright. I...wondered how your mood was on the journey and hoped you had a good dinner somewhere (though I doubt it) and that the sudden expulsion from the garden of Eden was not too painful for you. The garden was certainly enchanted. I wonder if it could be found again. It seemed to be waiting for us rather wistfully. The Primroses had so obviously bloomed their very best for us and had nearly given us up. But there was a promise for the future in the chestnut buds. The green fingers unfolding to show the potential flower seemed to say—come back in May and you will find an abundance of blossoms. We will go back, the garden needs us...

For a week or two all London has been our playground, the river and the parks and the streets in sunlight and moonlight and rain and even in the Strand which is horrid we came upon a little shrine at every few steps in the rainy darkness. And now we must work. It is good to take up a challenge. "Poete prend la plume, et ne donne pas un baiser, jusqu'au mois de Mai." ["Poet, take up your pen, and don't give away a kiss until the month of May."] The rose has blossomed out and is a pleasure to behold and conjures up looks, moments, words and the image of the giver.

...you can burn this letter and like the phoenix another will soon rise from its ashes. I know you are busy and must concentrate but I think I cant wait for a note so please write tomorrow and post it in time for it to reach me on Saturday. I will watch the posts. I think you wont disappoint me.

It is the middle of the afternoon, just 24 hours since we were in our garden and all nature smiled on us so sweetly and we smiled too right into each others faces vaguely remembering something we had shared in a long past age and had found in each other again.

Thank you for your smile . . . and for the things you said and the ones you didn't say and for everything else.

When her sister Helen married again there was, as it were, a vacancy at 272 King's Road. Into this unofficial post stepped the small and smartly dressed Beth Lipkin, a young Canadian musician who had arrived in Martigues while touring Europe before the war and got to know the family. She remained, in one dwelling or another, for the rest of Kathleen's life. Opinions about Beth Lipkin differ, although everyone agrees about her unwavering devotion to Kathleen. One friend described her as a schoolgirl with a crush on the head girl. Jackie Epstein called her "a definite man-hater" and said that she had kept photographs of Kathleen naked on display in her bathroom. Peggy Jean is alleged to have said that she was sure Kathleen and Beth were lovers. Others thought so, too, although no one very close to Kathleen believed that they were. (Jackie did not think so, although he believed that Kathleen had had some sort of flirtation with Pandit Nehru when he sat for Epstein in 1946.) Nevertheless, Beth's adoration for Kathleen could be unnerving.

Epstein's biographer Stephen Gardiner never stood with his back to Beth when they were in the same room, "Literally. I thought she might kill me." Esther called Beth "the Silent Witness." Epstein tolerated her. Helen couldn't stand her. There were those who thought that she was to Kathleen as Miss Tony Thomas was to Marjorie Garman, an indispensable aid and companion. To others, she had a more sinister quality, redolent of Mrs. Danvers in Daphne du Maurier's *Rebecca*. Roland Joffe recalled Beth's love for Kathleen as imprisoning. Someone once asked Helen why her sister put up with Beth, to which Helen answered, "You see, it's all about the letter," before clamping her hand over her mouth and refusing to say more.[*]

Kathleen was a strong, even dominant, personality. It seems highly unlikely that she would have allowed herself to fall into living under the same roof as someone who was effectively blackmailing her. More likely, Beth Lipkin started as a lodger who shared her deep interest in music, and simply stuck. She did receive handouts from Kathleen, and she performed certain tasks in return, like booking railway tickets and arranging journeys, but it was not a formal employment. Later on, one of Kathleen's granddaughters was especially attached to Beth, which deepened her connection to the family. Kathleen was evidently fond of her, as her many letters attest.[†] Beth became Kathleen's shadow.

Margaret Epstein died in March 1947. Returning from a day at Epping with her family, she fell on the steep steps of the

[*] A mystery surrounds this letter. See appendix.
[†] Beth Lipkin generously allowed her letters from Kathleen to be held at the New Art Gallery.

Kathleen at the beach during the 1940s.

house at Hyde Park Gate, fracturing her skull. She was rushed to the nearby St. Mary Abbots Hospital, where she died from a brain hemorrhage. Winston Churchill, who had moved into a house across the street from the Epsteins after his election defeat in 1945, wrote a kind letter of condolence. Peggy Jean, now married and living in America, came over for the funeral, and when she left, she took Jackie with her for a change of scene. This left Epstein alone in his sadly dilapidated house, having been looked after by Margaret for forty years. His wife had accumulated an entire room full of fusty old furs and trunks full of rotten clothes, all stinking of mothballs and cats. The rest of the house was cluttered and dingy.

This chaos suited Kathleen, who spent a great deal of energy on setting Hyde Park Gate to rights, painting the whole place white and emptying out the dross. "Kathleen was a good organiser, which I don't think Margaret was," says Jackie Epstein. "To an extent she'd lean on Beth to do mundane things, like order groceries. But it was Kathleen who went through the house like a dose of salts and opened it all up. Before her it had been in a terrible state, dark and dirty. It was a remarkable transformation: nothing had been altered structurally, but it was transformed. After Kathleen it was wholly unrecognisable."

But still Kathleen did not move in, much as Epstein wanted her to. Kitty was now married, and Esther and Theo had come to live at 272 King's Road after spending the war years with their grandmother. Esther was at school in Fulham. It was clear, by then, that Theo was suffering from some sort of mental illness. He was agitated, rambling, and suspicious. Kathleen spent much of her time looking after him. He had been a conscientious objector during the war, working on a farm at Harting. For some of these years, he had lodged with a family in the village and then lived alone in a caravan. He worked long hours as a cowman, painted, and took a correspondence course. His sister Kitty feels that the rigor of this schedule, which he adhered to, in all likelihood, to please his mother, was too much for him and dates the onset of his illness to this time. He had been a very gifted child, winning school prizes for literature and history, and singing in the choir. He remained a passionate reader and knew a tremendous amount about the history of art. Everyone who knew him loved Theo: he was gentle and kind, utterly without malice, although he did give people nicknames: Laurie Lee was "the poet" and Lucian Freud "Luxy." (Freud was, apparently, the only person not to have liked Theo and is said to have taunted him. On a trip to Paris, Theo's companion saw Freud across the street and pointed him out, whereupon Theo simply vanished in fright for the rest of the day.)

Friends from the time remember his nickname was "Sunny," which was apt, for Theo had a very amiable disposition. In fact, the nickname was "Sonny": Epstein always kept his New York accent, and the Americanized nickname may have been his way of acknowledging his connection to the boy. (Theo had his mother's surname, as illegitimate children then did.) For his part, Theo referred to his father as "the old gentleman." Friends remember them as being very similar. "He was very like his father in many ways, except Theo was mad and Epstein wasn't," recalls Stephen Gardiner. "He was always strange. He'd look at the house across the road—Argyle House in the King's Road— and say there was a man over there who was persecuting him." Relations between father and son had not been easy or straightforward, Theo having spent so little of his childhood with Epstein, but the older man was delighted when Theo began to paint, because it created a connection between them.*

Once installed back in London, Theo got to work. There was

---

* Epstein had wanted his son Jackie to become a painter and made him produce a picture each week for his inspection. Jackie dreaded this task, which was not optional. In time, he became an engineer.

*Theo and Kathleen in the garden
of 272 King's Road, early 1950s.*

a wild, tangly garden at the back of 272 King's Road, with a
shed at the bottom. This became Theo's studio. Like his father,
he had a complete disregard for domestic comfort. He slept in
the kitchen, though no one can remember why. He had by then
been diagnosed a schizophrenic, and he suffered from bouts of
paranoia. Still, he worked almost as obsessively as Epstein and
by 1950 had amassed enough work for an exhibition. Matthew
Smith wrote a foreword to the catalogue. "About the painting

of Theodore Garman, I can only say that I look at them with wonder, admiration, and even astonishment," he wrote. When Kenneth Clark went to Hyde Park Gate to visit Epstein, he saw some of Theo's work and exclaimed, "Why have I not been told about this wonderful artist?"

Theo had many friends, some of whom were also close to his mother. There was John Lade, a musician who worked at the BBC, and the architect Stephen Gardiner, who later wrote a biography of Epstein. There was also a small group of young Irish people, one of whom became a lodger upstairs at 272. One of the Irish was a doctor, Desmond O'Connor, whose wife looked like Rita Hayworth. Some friends of the time share a painful memory of being invited to tea by Theo. Ever genial, he handed around a plate of sandwiches, but in his muddled state he had filled the bread with green shaving soap. "He was completely oblivious," they recall. Another young married couple, Frank Ward and Kathleen Walne, both very gifted painters, had known Theo since childhood and lived nearby, in Elm Park Gardens. "He was an eccentric, but a lovely one. He was a favourite, in our household," recalls Kathleen Walne. "Frank and Theo had a great bond because they both ate and slept and drank painting. He used to come over to us, always hungry. He was very fond of cheese and we'd say, 'Have a piece of cheese, Theo,' and he'd take the lid off the cheese dish and eat the whole lot."

Although Kathleen Walne liked Kathleen Garman very much, she felt that the rarefied intellectual atmosphere at home may have been a strain on Theo. "She devoted all her time to Theo, to making him do well. They never had newspapers or magazines that ordinary folk read," she recalls. "That's why he liked to come to us, because he could put his feet up and read the paper and get something to eat." Her daughter felt that it was almost as if Theo and Esther weren't allowed to be children. Kathleen Epstein was extremely nice and kind to Hilary Ward, but she did have an aloofness about her, a coldness. "Her bearing was very regal. When she spoke to you she'd put her head on one side and look right into your eyes," she remembers. A nephew recalls this look as being frightening, almost hypnotizing, like a snake.

It was in the late 1940s that Kathleen made friends with Dr. Valerie Cowie and her husband, Dr. John Cowie, who were both highly qualified psychiatrists. By then, Theo was so ill that he needed extended treatment. He had become an inpatient at Northampton, in the same hospital where the poet John Clare was long incarcerated. Hoping for a second opinion, Kathleen asked Valerie to go with her to see Theo there. Dr.—now Professor—Valerie Cowie recalled, "She got a driver called Mr. Dean to take us and arranged a lovely picnic on the way. We stopped in this beautiful wood and she picked a lot of ferns and made plates from them, and there was French cheese and lots of fruit." This small anecdote is immensely revealing. A different kind of person, going to visit her only son in a mental institution, would have been too concerned, too preoccupied to stop for a picnic on the way. But such behavior was typical of Kathleen. She had to put a brave face on things, to be gracious at all costs. Anyone might have packed a sandwich or two for the journey, but only Kathleen would have stage-managed a picnic in an idyllic setting with such easy languor, going to the trouble of picking ferns for plates. It wasn't enough to create a good impression; people had to be enchanted by her. After meeting Theo, Dr. Cowie and her husband both informally confirmed the diagnosis of schizophrenia. "Theo was a lovely person, like a great bear. He was really warm hearted. But his illness made him very unpredictable. It was very difficult for Kathleen. She was always anxious to keep him entertained."

All of these friends helped to keep an eye on Theo, enabling Kathleen to spend time with Epstein. But in a letter to Peggy Jean, he complained of Kathleen's absence. "All this isn't good for me who needs looking after. The hundred and one things of practical life I am as little prepared to trouble with as anybody else. Kathleen's boy Theo gets worse and worse, all treatments have been tried with no results." Epstein needed help, as the re-

Jacob Epstein in his studio at Hyde Park Gate with figures of Kathleen. The second portrait of Kathleen (1922) is in the foreground, the fifth (1935) in the rear right.

markable self-absorption of this letter demonstrates. "He couldn't boil an egg," recalls Jackie. In time, Kathleen moved into Hyde Park Gate to be with him. It was then that he began work on his seventh, and final, bust of her. On Saturdays, they held musical evenings, inviting young art and music students (the future sculptor Lynn Chadwick was one), as well as longer-established friends.

Their daughter Esther joined them at Hyde Park Gate, to attend the nearby Royal College of Music. But it was not long before she met a journalist called Mark Joffe and moved in with him, just around the corner, at 18 Queen's Gate. Esther was beautiful, "like a lily, very elegant," recalls Rosie Price, the adopted daughter of Ada Newbould, with whom Esther had spent so much of her childhood. "She wore a lot of velvet, in dramatic black and claret colours, her skirts were longer than was then usual—ankle length—and she wore little fitted jackets with tight cuffs. Her hair was very, very black—coal black—and very long; at least to her waist." Esther spoke softly, with an almost childlike voice and was animated in short bursts. She was kind and given to tears. "Everyone who knew Esther was a little in love with her," recalls John Lade. She went to concerts by herself and cried when the music moved her. The young Wayland Young fell in love with her after spotting her at the Proms when, during the slow movement of Brahms's Piano Concerto in D Minor, he glanced across the audience and saw her weeping into a little handkerchief.

Mark Joffe had come to England from Germany as a refugee, and he had a young son, Roland. The boy's mother was Italian and had apparently married only to be allowed to stay in the

British Isles; she did not stay to see her son grow up but went to run a restaurant in Scotland. The little boy was in a temporary home in Kent. As soon as she heard of this, Kathleen hired a car for Esther to go to Tunbridge Wells to fetch him, but the child was so traumatized that when he arrived in London he refused to come into the house, but clung to the railings outside.

One or two family friends from those years deeply disliked Mark Joffe and could not believe that Esther would have moved in with him had it not been for her devotion to Roland. "Mark Joffe was a very demanding man," remembers Valerie Cowie. "We all thought she was throwing herself away on him. But she loved him and she made a lovely little nest right at the top of this tall house, with plants everywhere and spindly gilt chairs. She had a big bird-cage with a plant hanging out of it. She had a beautiful Venetian glass mirror and she used to give us Russian tea out of long glasses." To Kitty "he was very ugly: a small, unprepossessing man. I could hardly bear it, because this lovely girl...she had great pity for deprived people. She loved him. They had a dog. She loved making cakes." He was certainly very possessive and difficult, refusing to come out of the bedroom when her friends called, eventually forbidding such visits entirely. Kathleen kept a pragmatic silence.

The summers of the early 1950s were spent going back and forth from Italy. Kathleen had bought a house at Torri del Benaco, by Lake Garda, adjoining the one where her sister Helen was now living with her Italian husband and new baby. (Helen's was divided in two, and the lower half was sometimes let: Aaron Copland took it one summer but left in high dudgeon when his

*Jacob Epstein and Kathleen (right of photo), with Mark Joffe
and Esther on the left, circa 1948. Epstein's daughter
Peggy Jean is in the front on the far left.*

piano wouldn't fit through the doors.) Kathleen's half had previously been occupied by a woman and her goats, so another of her whirlwind transformations was required to make it habitable. Esther and Theo and Helen's daughter Kathy and Roland Joffe and all their young friends spent happy holidays here, swimming in the lake and going into the little town to get a taste of café life. They all traveled to Paris by boat and train, then took the sleeper into Italy. Kathleen often made friends with fellow travelers, especially children. Roland Joffe remembers endless games of whist on the train. But traveling with Kathleen wasn't always such fun. Kitty recalls long, uncomfortable journeys in third class, toward the end of which her mother would buy armfuls of lilies so as to look glamorous when she stepped off the train.

In 1952, Theo had a second exhibition at the Redfern
Gallery. He worked constantly, also studying his large collection
of books about painting, as well as reading widely: Keats's letters, Edmund Wilson's *To the Finland Station*, Thomas Mann,
the letters of Marcel Proust. Kathleen came over to 272 from
Hyde Park Gate to make sure he was eating because he often
forgot about food. Theo had mixed feelings about his mother's
help. One day, he was having a drink at the Six Bells pub with
a friend when Kathleen came in to look for him. When she left,
Theo said, "Thank God she's gone." She made sure that someone was always keeping an eye on him. John Lade, who was devoted to him, went to see him one day, and upon his arrival at
272 found smoke pouring out of the building and fire engines
in attendance. Theo wasn't there, but when he turned up, he ig-
nored the blaze and the firefighters, ambling over to John Lade
with the words, "Oh, there you are," as if nothing was wrong.
With the Lades he went on trips to Paris and to Italy, always
traveling in his thick tweed coat, whatever the season. But both
Parisian visits had ended abruptly with Theo going back to
London alone, because "he didn't like the police." Yet however
eccentric he seemed, everyone was fond of him, drawn to his in-
nate sweetness. Students from Chelsea Art School, which was all
but next door to 272, came to see him in his garden room and
enjoyed his company. But in January 1954, after a struggle with
the ambulance men who had been called in by Kathleen, he
died, apparently from a heart attack. He was twenty-nine.*

Epstein's knighthood (recommended by his friend and neigh-

* For a more detailed analysis of his death, see appendix.

bor Sir Winston Churchill) had been announced in the New Year's Honors of 1954, but Kathleen was too distressed by her son's death to attend the investiture at Buckingham Palace in February. Esther went in her place. A photograph of father and daughter that day shows Esther looking drawn, with dark circles under her eyes.

Esther was devastated by her brother's death and also had difficulties of her own. She longed to extricate herself from her troubled association with Mark Joffe but didn't want to leave little Roland. For a time she moved out and went to stay with her parents, taking the boy with her. Then, at one of a number of parties given in St. George's Square by Phillip Toynbee, Jocelyn Baines, and Ben Nicolson, she met a student called Mike Rutherston,* who shared a top-floor flat with them. Rutherston, a fledgling cartoonist very much in the manner of his godfather Max Beerbohm, fell in love with her on sight and asked her to marry him. He was fair and handsome, with thick swept-back hair, but he was wild and terribly neurotic. When Esther turned him down, he killed himself by putting his head in the oven and turning on the gas.

Kathleen's psychiatrist friends, John and Valerie Cowie, took Esther away with them to France in April 1954. She would get up very early in the morning and go for long walks by herself, and John Cowie confided to his wife he "didn't like the look of it at all," meaning that he could see the symptoms of an incipient depression in Esther. She had hardly been back in England for any time before she left again, this time to stay with friends in Italy. They remember her crying a lot, but they were not un-

---

* Mike Rutherston was the son of the artist Albert Rutherston. He was a cousin of Epstein's old acquaintance Sir John Rothenstein, a director of the Tate Gallery.

duly concerned. Esther did tend to weep, and they did not suspect there was anything seriously wrong. But back in England in the late autumn, it became obvious that her increasing unhappiness had become very serious when she attempted suicide by taking an overdose of sleeping pills. Valerie Cowie found her in time to call an ambulance, and she was saved. Kathleen found a suicide note torn into little shreds and pieced it together.

John Cowie believed that Esther's situation was now so grave that he advised ECT (electroconvulsive therapy, the only course of treatment for severe depression which can take effect immediately), but the young registrar at the hospital where she had been admitted did not believe that this would help her. Instead, he advised a course of psychotherapy and soon discharged her. There were those who believed that she should not have been left alone after leaving the hospital, but Kathleen thought that it would do her good to have her independence and installed her in a small flat in Pembroke Villas. She was to start a job, working in a bookshop with a friend.

Esther and Roland used to make jellies together, and they had made a red jelly in a rabbit-shaped mold, which they'd turned carefully onto a plate and put on top of the refrigerator, to have later. But it never got eaten. In November 1954, alone in her new room, Esther committed suicide by gassing herself. Roland Joffe thinks he remembers looking at the jelly, as it grew a green hair of mold, and somehow apprehending through this that the spirit had left the home, that Esther had gone and was never coming back. But it was not the custom of the time to inform small children about death, and he was told that she had gone to Italy.

Esther was buried with Theo by the wall in the churchyard at South Harting. Three of the four children Kathleen had borne had now died. (A baby, born after Esther, had been the victim of a cot death in infancy. Kathleen told Valerie Cowie that she had been playing the piano in the same room and when the piece came to an end she turned around to look into the crib, and the baby had stopped breathing.)

Valerie Cowie went to sit with Kathleen on the night that Esther died. She remembers that Kathleen sat rocking back and forth with grief. She had been doing a piece of knitting, and she said, "I wondered what this knitting was for and now I know: it's for wrapping around Essie's feet." Epstein talked about a new block of marble that had just arrived and his plans for the stone: he didn't, or couldn't, mention his daughter. Soon afterward, however, he wrote to Peggy Jean in America, "Esther who was so beautiful and well-loved by all committed suicide. She it seems was much troubled first by Theo's death nearly a year ago and then by the suicide of a young student friend . . . Esther had no mind to marry altho' she was courted by several young men. She attempted to do away with herself about six weeks ago and was saved by intensive treatment: she was three days in a coma. After that she was treated in a home where she seemed to recover her peace of mind . . ."

It was a great sadness to Kathleen that Esther had never had a child of her own, but Roland was treated exactly as if he had been her son. Not long after Esther's death, Mark Joffe disappeared to Paris, leaving the boy in Kathleen's care. Roland called Kathleen Mum because that's how Esther had referred to her. On Sunday evenings, Epstein would read aloud to Kathleen and Roland, usually from Charles Dickens. "I think my love of stories came from those evenings," says Joffe, now a successful filmmaker: "I used to make pictures in my head while he read." He also believes that he got his own love of women from Epstein. "I realised they must be pretty fantastic, for him to be so rapt by them. And I think I got my model of what women were like from Kathleen and Helen: witty and strong and beautiful and loving to argue. Helen was the most extrovert, Kathleen the strongest character, although Mary came close. They were *significant* women." Later, he was sent to Carmel College (they always called it Caramel College), the only Jewish boarding school in Britain, at Kathleen's expense.

Months after Esther's death, at one of the musical afternoons which Kathleen and Epstein sometimes held at Hyde Park Gate, the visiting pianist played Schubert's *Death and the Maiden*. It had been a favorite piece of Esther's (Lorna, too, loved the music and made many paintings on the theme). During the recital, the doorknob turned and the door opened, but no one was there. Kathleen reported this to Valerie Cowie, clearly convinced that her daughter's spirit had opened the door. "Kathleen believed very much in that sort of thing," says Dr. Cowie. "She was quite sure she was psychic."

Perhaps such a belief afforded her some comfort from the overwhelming grief of losing two children within a year. Epstein was worried that she might collapse, writing to Peggy Jean, "Kathleen is terribly upset . . . At the moment she can think of nothing else and lives in a distracted and almost demented manner ... I hope Kathleen will come out of it, in time. Don't write to her, as that would carry it on and I am hoping that the days will have its effect on her: altho she even now misses the memory of Theo, visiting his grave in the country frequently." On November 27, he wrote to her again: "Things are still bad ... I hope time will heal so much tragedy." Kathleen kept her sorrow to herself and didn't abandon her pleasures altogether. Her letters attest that she still loved music and painting, books and travel. "She had this tremendous zest for being alive," says Roland. "She had an immense capacity for turning everything into an adventure."

The next year, in June 1955, Epstein and Kathleen were married in a small private ceremony at the Fulham registry office. He wrote to his daughter Peggy Jean, "Yesterday Kathleen and I married. I had wanted this for a long time but Kathleen could not stand the publicity and so we had to resort to a ruse which completely dished all journalists and photographers and the event passed off quietly without the least fuss. Our odd position has for a long time been a source of embarrassment and now I hope things will be better ..." (By July, the *Daily Mail* had found out about the marriage and printed a headline: "Epstein Weds His Model: Secret Out!") The newlyweds had known each other for more than thirty years, and though Epstein was in his midseventies, he remained as much in love with Kathleen as ever. "How wonderful to get your letter addressed to me as your husband," he wrote her in July, "and to think of you as my wife." He called her "sweetheart darling." Now she was Lady Epstein.

Chapter Twelve

OUR WINTER SEASON

During the winter of 1933–4, Mary and Roy Campbell, unconventional as ever, made their home in the prostitutes' quarter of Barcelona. Late at night, while the children slept, they would go out to hear the Gypsies sing and watch them dance flamenco. The Aldous Huxleys, accompanied by Sybille Bedford, took them out to dinner one evening. Mary enrolled at the art school, where she befriended a group of impoverished Jewish refugees from Germany whom she invited back to the apartment for meals.

Tiring of Barcelona in March 1934, the Campbells traveled by train toward Valencia, in search of an agreeable village where they could rent a house. Their failure to secure lodgings led to Campbell drinking more heavily than usual, which in turn led to a quarrel that lasted for several days. The children always remembered this row, and that in the course of it Mary announced she was leaving him. But in the end, they were cheered up by a small win on the lottery and decided to stay together.

At Altea, near Alicante, they found the perfect place to live. Set between the mountains and the sea, the village was unspoilt, with small whitewashed houses running up to a large, blue-domed church, and orange groves that sloped down to immense fields of blue artichokes. Here Mary secured a white farmhouse covered in heliotrope and jasmine, with a traditional outside bread oven and a flower bed full of roses and geraniums. They took on a pair of maids. The Campbells had, their daughter Anna remembered, a "deep horror" of tourists, so they were glad there were only Spaniards in the vicinity. Mary nevertheless considered herself superior to the natives, and "always took it for granted that anyone south of Calais was a nigger. She felt this instinctively and didn't have to say it."

Now that their parents were settled, the children were left to their own devices once again. Tess, who was eleven, decided that she and seven-year-old Anna should attend the village school, which involved a long walk each way. Anna remembered it as being ten kilometers (about 6.2 miles). Mary and her husband, however, took their daily walks alone, having decided that their daily perambulation was too arduous for children. The girls were heartbroken when their pet dog died, apparently of starvation, after getting a fur ball lodged in her throat. Neither adult noticed that the dog had been vomiting all her food. Once again, the girls contracted head lice, which Anna remembered as "great fat slugs which ran over the scurf comb in thousands." Perhaps their mother did not see them scratching their heads, since her husband had forbidden her from wearing dark glasses because they smacked of the dreaded tourist. Forever squinting against the Spanish sun and the glare of the whitewashed houses, she is said to have injured her eyes irreparably during this time.

It was at Altea that the Campbells had their religious awakening. They were impressed and moved by the unquestioning faith of the community and began to study the Spanish mystics. They read the life of Saint Teresa of Ávila. Mary came to believe that suffering was to be embraced instead of avoided, that suffering would strengthen the soul. She decided to become a Roman Catholic and take instruction from the local priest, Don Gregario. (This priest was shot dead by Republicans during the Spanish Civil War, along with many of his parishioners.) Campbell soon followed her into the church, taking Ignatius Loyola as his Catholic name, while Mary somewhat unimaginatively took Mary Magdalen as hers. They made new marriage vows, and at the beginning of Lent 1935, began to attend mass. Until then, Mary had always said that she lived purely for sensation, confident that life

Mary, circa 1938.

would provide one excitement after another. Subsequent to their conversion, hedonism was replaced by an exaggerated piety.

Mary's devotion to the Catholic church was so great that no

one was spared her eulogizing. Her sisters tired of it, and Kathleen's son, Theodore, nicknamed her "Popey." Mary also liked the accoutrements of faith. When he had visited the Campbells in Spain, Laurie Lee had noted the crucifixes in the house and Mary carrying roses to church. He had observed in her "the banked-up voluptuousness of a young and beautiful convert, holding to this single passion in which all hungers were answered and all doubts quietly put away. Here, romantic love was put on ice, sealed by an unfaltering spiritual flame, and accompanied by a vocabulary of torment, physical denial and ecstacy which promised an eternity of sensuous reward." Listening to her, Lee had felt himself pulled by the "seductive faith." "She sat among her pin-up icons, smiling quietly, unshakably contained. 'Don't you see?' she kept saying (we were damned if we didn't). 'You can't *imagine* the utter peace.' "

The restless Campbells moved again in June 1935, this time to Toledo. Once more, Mary found a lovely place to live. Formerly the summer home of a cardinal, it was a tall and ancient house hidden within Moorish battlements. It had a garden with a fountain and roses, green shutters, cool stone walls, polished wooden floors, and dark Spanish furniture. (This house, the Hostal del Cardenal, later became a luxury hotel.) They installed horses in the stables and took on two maids. Mary used the ancient cistern for cold storage, lowering a basket filled with wine bottles into it, as well as keeping fruit and salads there. It was in this house that Laurie Lee stayed, sleeping on a mattress propped up with books, before he met Lorna.

Roy worked on his book *Mithraic Emblems*. When she wasn't in church, Mary rode into the surrounding countryside on one of the horses to draw and paint, bringing huge bunches of wildflowers home with her. Her faith was now so consuming that she recited the daily office, attended mass every morning at seven o'clock, and afterward went to Benediction. She soon volunteered to become a Carmelite tertiary, which meant giving up makeup and jewelry and wearing nothing but brown. She began to wear a scapular (two strips of cloth which hang over the breast and back, joined at the shoulders) and fasted at least twice a week. With Tess attending school in England for a time, Anna often accompanied her mother to mass, and Mary was thrilled by the child's devotion. In fact, Anna was more lonely than pious, and going to church was the only guaranteed way of being with her mother.

Whenever Mary went out without her daughter, she always gave a time by which she would return, but she never kept to it. Hearing the many bells that rang over the city, Anna would think they were sounding for her mother, who must have died in some terrible accident. Eventually, the maids were alerted to the child's distress, and Anna was sent to school. "Because I was now speaking Spanish all day away from home, Mary always made me read English aloud to her in the evenings, to see that I kept her very pure accent," wrote Anna. In other respects, Mary was less punctilious. Anna became ill with typhus, believed to have been contracted from body lice.* Mary got the maids to rub the child all over with foul-smelling sulphuric cream to kill the lice, but forgot to see about rinsing it off. When Mrs. Garman brought Tess home to Toledo, Anna felt terribly humiliated to greet her loved ones, still covered in this stinking unguent. "You smell terrible!" Tess told her, by way of hello. Anna promptly burst into tears.

---

* Typhus fever may be borne by fleas, mites, or lice.

Tess now remained with her parents and began to attend school with her sister. In her spare time, she practiced her embroidery, becoming an expert needleworker. "Everything Tess took up she did brilliantly," Anna said. "She would get on a horse and gallop off quite fearlessly, [while] it had taken me a month to learn to ride ... but I was prettier, so Mary spoiled me. She showed plainly that she loved me more." Mary's preference for Anna may well have been a kind of extended narcissism: Anna looked very much like her mother, while Tess was more like Roy in appearance. At this time, Tess had reservations about Catholicism, having scruples about going to Holy Communion. "She could not imagine being perfect enough to receive Christ," said her sister. She began to have fainting fits. Then, at fifteen, "she became crazed with anorexia nervosa." She would put her food onto Anna's plate. At school, the other children would chant: *"Anita, Anita, no comes croquetas que tu pobre hermanita se queda esqueleta!"* ("Annie, Annie, don't eat croquettes; your poor little sister is turning into a skeleton.") Her illness, incorrectly diagnosed, was to last for many years.

The Campbells could not have picked a worse time or place to become ardent Roman Catholics. Civil war was about to engulf Spain, and in March 1936, Toledo erupted in riots. The church became the target of the socialists and anarchists: churches were set on fire and priests attacked in the streets. Mary and Roy gave sanctuary to several monks. These monks wore lay clothes, but even so it was an extremely dangerous undertaking to house them. In July, seventeen Carmelite monks were shot dead and left to rot where they fell. The Campbells also took in many of the Carmelites' most precious papers for safekeeping, including the personal papers of Saint John of the Cross.

One day, a little girl who used to throw stones at the Campbells when they were out riding arrived, breathless, to say that they must take down the Union Jack flag which they had hoisted onto a tree to declare their neutrality. The little girl told them they must fly a red flag in order to escape attack. "But we have nothing red in the house except my silk pyjamas!" cried Mary. "We'll fly those, kid," replied Roy. But the pajamas wouldn't cling to the branches. Helpless with laughter, they watched the makeshift red flag slither again and again to the ground. A few days later, the house was searched by militiamen. Although they laid their rifle butts on the very chest that contained the Carmelite papers, they never looked inside. A copy of Dante caused a moment of panic: "Italian! Fascist!" shouted a militiaman. Luckily, Roy was able to show them his Russian novels in mitigation. "When I was examined by the Red chief of police, my wife rushed in, lovely, furious, and ready, if need be, to die," wrote Roy.

Evidently, they had to leave Spain. There were daily rifle battles in the streets, they had no money, and they were living on cucumbers from the garden. They spent their days huddled inside the house. Rescue came in the form of a man named Angel, whom Campbell had helped to write some poems. He turned up at the house carrying a dead child, killed by a shell, and a paper bag full of money, which he gave to Campbell. By bribing some militiamen, the family was able to get out of the city in the back of a truck used for carrying the dead. They made their way to Madrid and then to Valencia, where they were given passage on a refugee ship bound for Marseille. Others were fleeing Spain with them. Laura Riding and Robert Graves were also on board, and the Campbells soon befriended them, traveling via Paris to London together.

Back in England, the Wisharts temporarily put them up in Binsted. Mary moved to a cottage near Petersfield when Campbell enlisted with the King's African Rifles, with which company he was posted near Nairobi, in 1943. Wartime meant that the checks he dispatched to her were often lost or late, and she was obliged to take a job as a cook, and Tess as a chambermaid, to a rich widow, Mrs. Bridson. Mary had done almost no domestic chores in her life, having been able to employ French and Spanish maids for very lit-tle money, and she felt humiliated by her new circumstances.

Mary moved to London in due course, working for a French officer in Mayfair. Roy joined her in 1944 when he came back to England no longer fit for active service, having suffered from malaria and sciatica. Reunited, the Campbells settled in London, at 17 Campden Grove in Kensington. Tess had enlisted in the Women's Royal Naval Service, but had been deemed physically unfit. She was extremely close to her father, and his prolonged absence with the King's African Rifles upset her. A flying bomb that destroyed the house next door was too much for her shattered nerves. She was treated in various hospitals and clinics, but in 1948, she attempted suicide. Her health was now so poor that she had to be restrained and kept under continual observation. Life was also difficult for her sister, Anna, who was trying to become a dancer. "I never became a really professional dancer because Mary curtailed my lessons," she remembered. "Mary said she could no longer pay for my classes, so I went through much suffering...she found that life was easier for her when I toured the U.K. instead of having to pay for my training. This led to years of frustration on my part. Because I was graceful and dainty, people who auditioned me took it for granted that I must have good technique, but they were later disappointed." When Anna was offered the opportunity of touring India with the Anglo-Polish Ballet, Mary refused to allow her to go. "Why did I allow Mary to make destructive decisions for me? She had a very strong will and she used blackmail very subtly so that I was trapped. If keeping Mary happy was the point of my existence I certainly carried it out. But I had to live, too...she thought, because I was intelligent that I could get away with anything in life. She never, until I had a breakdown [in 1948], realised how fragile I was." Her cousin Kitty also remembers how harsh Mary could be. "Those poor little Campbell girls—what they endured at the hands of their parents—their Scottish blood must have supported them...I do have to say that the three most alluring and striking of the Garman sisters, Mary, Kathleen, and Lorna, *were* sadistic toward their children, always streaked with a deep love—a bad mixture."

Roy Campbell was chaotic, often drunk, sometimes pugilistic, but not unkind. He took a job working for the War Damage Commission in order to pay Tess's medical bills. He was also trying to get his latest poems into print, and T. S. Eliot, then director of publishers Faber & Faber, became a friend. Contentious as ever, Campbell wanted to include a preface highly critical of Louis MacNeice, Stephen Spender, W. H. Auden, and Cecil Day-Lewis—"MacSpaunday," as he called them. He was persuaded to drop the preface, but he maintained his contempt for them all and was furious when Spender denounced him as a fascist sympathizer. He took his revenge in April 1949, at a reading of Spender's in Bayswater, when he landed him a right hook on the nose. Louis MacNeice later gave him a nosebleed when they exchanged blows, but the two subsequently became friends. He quarreled also with Geoffrey Grigson, threatening him with his walking stick until the younger poet fled into the safety of a nearby cake shop.

By now, Campbell had a job with the BBC, producing talks for the Third Programme (now Radio 3). He also made friends. He renewed his acquaintance with C. S. Lewis and met J. R. R. Tolkien (Campbell reminded him of a character of his own invention, Aragorn, from *Lord of the Rings*), both of whom he visited in Oxford. He and Mary also spent many evenings with Dylan and Caitlin Thomas, Edith Sitwell, and William Walton. Evelyn Waugh came to dinner. One St. David's Day, Campbell and Dylan Thomas ate a whole bunch of daffodils in the pub, for a bet. When Edith Sitwell became a Roman Catholic, she was much influenced by Mary. "My mother could talk very, very beautifully about those things and Edith became fascinated," recalled Anna. On being received into the church, Sitwell asked both the Campbells to be her godparents. Another new friend was the sculptor Hugh Olaff de Wet, who made bronze portraits of both the Campbells.

Mary and Roy became actively involved in a new literary magazine, *Catacomb*, which was taking shape under the patronage of another younger friend, the poet Rob Lyle. Lyle was a recent convert to Catholicism (Roy was his godfather, too), and he admired and liked the Campbells. In time, he would marry their daughter Anna, but for the present he enlisted their help with *Catacomb*, persuading Campbell to become its paid editor, a job Mary helped with. It was here that much of Campbell's translation work appeared: Baudelaire, Rimbaud, Lorca, Apollinaire. It was at Lyle's suggestion, too, that he wrote a second volume of autobiography, *Light on a Dark Horse*. Aware that its contents would scandalize some readers, Roy called it "my autobuggeroffery."

Lyle also employed Tess. She sent out the subscriptions of *Catacomb* each month, and the methodical task helped her (the work cannot have been too onerous, since subscriptions never went above two hundred). She began to eat a little more. A young Polish doctor, the friend of a neighbor, came to see her and asked her a simple question: what did she most want in the world? "To go to France or Spain," she answered. This chimed with Roy's own desire to get back to the south, and in May 1950, father and daughter set out for the French Mediterranean, where Tess smiled for the first time in several years. She also began to put on some weight. It is interesting that the beginning of her recovery should have occurred while she was away from her mother. With Tess and Roy in France, Mary went to Italy, where she was joined by Anna, for an audience with the pope, the first of three such pilgrimages she was to make. Later in the summer of 1950, the family met up in France. Augustus John was finishing his autobiography, *Chiaroscuro*, while staying in a nearby village, and as in the old days at Martigues, the Campbells and the Johns saw a great deal of each other.

Back in London that autumn, *Catacomb* limped on. Though it published Charles Causley, Sacheverell Sitwell, John Heath-

Stubbs, and others, and received a fan letter from Ezra Pound, the magazine's combination of reactionary Catholicism and poetry did not encourage sales. Lyle decided to fold the magazine and, with Campbell, determined to find a small farm to rent, somewhere warm. Campbell had visited Sintra in Portugal during the late '30s. In March 1952, he went back there with Lyle, finding a small farm in the hills.

In the 1950s, Campbell toured America and Canada, giving well-paid lectures and readings. He particularly enjoyed meeting Theodore Roethke, who introduced him to bourbon whiskey. He was delighted, too, to sit next to Eleanor Roosevelt; "we almost fell for each other," he told Mary. He also went back to South Africa. In 1952, he had met Alan Paton, who found that Campbell's conversation lit up the lunch table like a flare. They had cosigned, with Laurens van der Post and Uys Krige, a letter to the South African government, protesting against plans to take mixed-race voters off the electoral roll. Campbell received an honorary degree from the University of Natal but missed Mary. "I am longing to see you beloved," he wrote, "but this is the last time I will ever set eyes on my beloved country (How I love it!) so let me take in all I can before I finish with it." But his wife was sick of his absence. "All I know is that I have had enough of being quite alone here, and I am *longing* for you to come back," she wrote. Mary was looking after the farm on her own, which was hard work. Rob Lyle had gone back to England, Anna had made her first marriage (to a Spanish nobleman), and Tess, too, had married. In the wedding photograph the expression on her pinched little face is remote, like a guest at the wedding of someone she hardly knows. The marriage did not last.

The Campbells had always thought of moving back to their beloved Spain, but an accident on their small farm made any such plans impossible. The laborer they employed on the farm fell from a ladder while he was pruning a tree. Although he had been only three feet above the ground, the fall was awkward, resulting in a fracture that a series of operations could not rectify, and he was left crippled. The Campbells had no insurance, and they were obliged by Portuguese law to pay the man's medical bills and two-thirds of his wages for the rest of his life. Never rich, they were left all but destitute. To raise money, Campbell embarked on another lecture tour of America, and they moved to a much smaller house in the nearby village of Linho. It was a modern property, and cramped. Only after they had inherited some money from Campbell's mother were they able to settle a lump sum on the injured worker, as well as buy a plot of land on which to build a house, the Casa da Serra. It was to be the first they had ever owned.

Campbell was happy, writing poems which still had Mary as their tutelary spirit:

> *But for the firelight on your face I would not change the sun*
> *Nor would I change a moment of our winter season, no,*
> *For our springtime with its orioles and roses long ago.*

Edith Sitwell, having read the second edition of his *Collected Poems*, wrote to him: "It is, I may say, one of the great things in my life, this lasting love between you and Mary, in an age of little, fireless unreal would-be emotions, gone in a moment, everything turned to ashes, to know that such greatness and undying love exists, unquenchable. You deserve, and she deserves, your poetry."

In April 1957, Roy and Mary drove to Spain for the Holy Week ceremonies in Seville, spending several days in Toledo, always their favorite place. On the journey back, with Mary driving, a front tire burst and their small car swerved into a tree. Forty years earlier, during the First World War, she had driven her father's car into a tree, but they had both escaped unscathed. This time the accident ended in tragedy: Roy died minutes later. He was only fifty-six.

Vita Sackville-West sent her condolences. Mary replied: "Thank you for your loving letter. I know that in spite of everything it is absolutely sincere and it comforted me. There is one thing I have always meant to ask you, although I am quite sure it is ridiculous and unnecessary—whether you still have any letters of mine? perhaps among old papers and forgotten? In any case will you be so very patient and kind as to tell me that you have not. It is rather a matter of conscience with me, so forgive me." Despite calling the "whole business" a "painful memory," Vita had not destroyed the letters; nor did she do so now, despite Mary's request. She did not return them, either. It would have been unlike her to get rid of such evidence testifying to how captivating she had been. Vita kept the letters, along with a journal Mary had written about their affair.

The Casa da Serra outside Sintra is reached by a steep and winding avenue of eucalyptus. The pink house on this hill of pines and cork trees had been the dream of both Campbells. As it was, Mary took up residence there alone. She was nearly sixty, but she looked much younger and was still beautiful. One visitor of her late years described her as having hair more dark than gray, which seemed full of electricity, standing up around her head as if alive. He found her striking, with her particularly fine bone structure and beautiful cheekbones. She had dark, vivacious eyes, and there was almost an impish wickedness about her. Anna says that Mary received numbers of proposals of marriage during the long years of her widowhood, but she turned them all down. Her last great love was for Pope Paul VI. In her final private audience with him, she presented him with a copy of Roy Campbell's *Collected Poems*. The prior of the Dominicans in Lisbon became a visitor, and she attended mass every morning.

Once she had recovered from the injuries of the car accident that had killed her husband, Mary began to make a garden. She planted red and white and palest pink camellias and mimosa. One summer, there were forest fires, and all the trees and shrubs on the hill went up in flames, forcing Mary to leave the house, which seemed certain to be razed. But the wind dropped and the house was saved. Mary began the arduous work of planting her garden anew.* During the early years at Casa da Serra, she was able to afford a chauffeur/gardener and a cook/housekeeper, but over time the income Roy had left her diminished, and by the end of her life she was penniless. De-

---

* Oddly enough, the widow of Campbell's old adversary Stephen Spender endured the same fate years later, at her house in Provence. Lady Spender courageously refused to leave the premises, obliging the fire brigade to attend the scene. As a result, her husband's library was spared the flames. She, too, replanted her garden.

spite the favorable climate, the house became uncomfortably cold, at least to visitors. Mary, who had never lived in a centrally heated house in her life, did not seem to notice. The furniture became rickety, the rugs threadbare. In time, Tess moved into the house with her mother, and the two of them had hardly enough money for food. "Neither of them seemed to care: certainly Mary didn't. She was almost indifferent to material things, didn't much notice what she ate," recalled a visitor. She was generous, giving away her pictures to people who said they liked them. Lorna's son Michael paid homage to her in his autobiography. "I cannot find words to pay my aunt Mary the tribute she deserves," he wrote, "she paints roses; listens to Mozart; prays. She is poor. That is nothing to her."

Mary listened constantly to Mozart on the old-fashioned gramophone in her studio. She especially loved the quartets, and in her last years she read Mozart's letters again and again. She painted occasionally, and she drew and sketched a great deal. But she was not precious about her painting: she called her studio her "workroom," and the floor was covered with discarded drawings and old letters and papers.

Her daughter Anna believed that Mary never fulfilled her own talents.

She was very musical. She accompanied herself on the guitar, and she could take up the piano after years of neglect owing to our never owning that instrument. She really painted divinely. These gifts were extraordinary. Such ease! [Earlier] she took to riding with the greatest facility. She managed to get on in foreign countries without any complications. She trained peasant girls to do housework in her place. Yes, Mary was highly gifted. But you cannot be everything. In order to be the wife of a vulnerable poet of genius, she often neglected her gifts in order to have the stamina to cope with a life of great hardship... She had to be tough to live with my father, but about life he was far more sensible than she was... she was a dreamy woman, utterly irresponsible. Nevertheless, she was the mainstay of the family.

Although they were reconciled before Mary's death, Anna and her mother fell out. "I had had a very passionate relationship with her; she was a very passionate woman in all her relationships, with men and women. But she was a raving Catholic and it was all so heavy... I was suffocated by my mother, always. Tess looked after her in her old age."

Peter Alexander, who wrote the first biography of Roy Campbell, stayed with Mary at Casa da Serra for two weeks or so in 1975 and again in 1978. He remembered,

Her voice was high, clear: I see I called it "silvery" in the biography, and I'd stick to that word. Rather "posh" English accent even after decades of living outside England; she seemed not to speak Portuguese except with difficulty. She wore long dresses, I mean mid-calf rather than ankle length but longer than was then fashionable, and cardigans with buttons and pockets into which she jammed her hands so that the cardigan was pulled down in two points: that's my vivid memory of first seeing her... though it is an unfashionable thing to say, [she was] intensely feminine, vivid, flirtatious even in old age, particularly when recalling her love affairs...I could see how easily she tolerated Campbell's bringing home drunks from bars and expecting her to feed them; she'd have done the same herself, I think...She was full of life and very clear-minded... she drank very cheap Portuguese white wine mixed with lemonade, describing the result as "better than champagne." I drank lots of it with her, and agreed after the second glass... after dinner she would talk again, and in this mellow state would remember her love affairs...she was gleeful about the "naughtiness" of these memories, I think, and I got the impression that she regretted nothing. She had enjoyed her beauty to the full. It was after dinner that she would bring out her portrait of one of her lovers, Uys Krige, in the nude.

"I don't know whether I've managed to convey the fact that I liked her a lot," he concluded. "She was a very warm and generous human being."

According to Anna, her mother lost her faith about two years before she died. It seemed to disappear suddenly and without reason, like a bird alighting from a branch. Nevertheless, when she suffered a stroke, the priest was summoned, and Mary made her last confession and received the final sacrament. Mary's maid told Anna, with Iberian piety, that after receiving extreme unction she had looked like a young bride. She died on February 27, 1979, and was buried with her husband in the cemetery of São Pedro, near Sintra.

Chapter Thirteen

DEALING WITH DREAMS

For Epstein and Kathleen, the final years of the 1950s were a happy time. Their daughter Kitty had had two daughters with Lucian Freud, and their grandparents loved the girls. In 1957, Kitty married again, the marriage with Freud having ended. Her new husband was an outstandingly handsome musician named Wynne Godley,* who became the model for Epstein's greatest public work, the huge bronze of Saint Michael at Coventry Cathedral. The Epsteins traveled to Venice and made a tour of French cathedrals, taking in Lake Como as well as a Holbein exhibition in Switzerland on the way home. Epstein had suffered a heart attack in March 1958, but was soon well enough to work again. "It is absolutely expedient for E to have another break, although he doesn't realise it," Kathleen told Beth Lipkin, concerned that, as usual, he was working too hard. During the hot summer of 1959, they en-

---

* Wynne Godley is now an economist of world renown.

joyed a trip to Paris, spending many hours in the Louvre and going on, once again, to Venice. Nevertheless, he wrote to his daughter Peggy Jean that he was "very little of a gadabout these days. I get very tired and I must reserve my energy for the large works I have in hand." Chief among these was the *Bowater House Group*, his last big commission, which stands today at the Edinburgh Gate entrance to Hyde Park, just off Knightsbridge.

In August, he and Kathleen went to see his dearest friend, Matthew Smith, who was ill, and drank Champagne together while they reminisced. They lent Smith a little landscape by Renoir to keep by his bedside. The next day, Epstein worked all day in his studio before dining with Kathleen and Beth Lipkin at nearby Ciccio's. When they got home, he sat outside looking at the stars, singing Schubert songs. Within an hour, he had had a fatal heart attack. "It is as if some of my world has crumbled away," T. S. Eliot wrote to Kathleen. "You have our warmest sympathy. We loved him." The Epsteins had been among the handful of close friends who attended the poet's seventieth birthday party, the year before, and it was Epstein who had proposed the toast, to which Eliot replied that it was the happiest birthday he had ever known. Kathleen placed the poems of Walt Whitman in Epstein's coffin.

Further sadness came for Kathleen six weeks later, when Matthew Smith died. At the funeral, Laurie Lee looked across at "The gaunt South American Indian cheekbones of Lady E—and little Kitty a gypsy beside her." Inscrutable as ever, Kathleen did not share her widow's grief. Instead, she became an exemplary dowager, putting her odd serenity and all but tireless energy to good use in the management of her husband's affairs. She made important donations of his work, including the gift of two hundred plaster casts to the Israel Museum, Jerusalem, and arranged exhibitions. Epstein had been the one person in Kathleen's life who had not followed her wishes; now she made it her task, as she always had, to follow his. Her friend Rosie Price found her "a very dramatic person and a very practical person, too; very well organised and efficient. She had presence and when she walked into a room she was magnetic ... But she devoted her life, really, to that man and his work. All Epstein wanted to do was work. Her life was almost a sacrifice to him, to his art."

In 1960, the Arts Council displayed his collection of tribal art in London's St. James's Square, while the Leicester Galleries showed a selection of his bronzes and drawings spanning fifty years. It was during that year, too, that Kathleen unveiled the bronze *Saint Michael and the Devil* at Coventry Cathedral. With Richard Buckle, she organized a major retrospective in Edinburgh in 1961, as well as a comprehensive book of Epstein's work. Sir John Rothenstein made the selection for a commemorative show at the Tate, also in 1961. "Daddy I know could not have found anybody better than you to look after his interests as you know how he wanted things seen to," Peggy Jean wrote to Kathleen. "I am sure as he rests it is in real peace." But despite all this work to secure her husband's memory, Kathleen came under fire when it was rumored that she had had extra castings made of some of his bronzes. This has always been denied, although one or two of her old acquaintances ventured in private that it was so. "It was only two," suggests Roland Joffe, "and she enjoyed it being rather a scandal; she liked the drama of it, relished it." When collectors came to Hyde Park Gate, she would artfully arrange flowers in a tall vase and light a fire in the viewing room. "You know the old formula," as she described it to Beth.

Is it surprising that, after an entire adult life flying in the face of convention, her newfound respectability should have given her a certain amount of pleasure? Roland Joffe attests that "she liked being a Lady and she enjoyed it that lots of people called her 'your ladyship.' But she preferred being called Kathleen. She wasn't a snob. She felt that everyone had a story as profound as everyone else's." Rosie Price agrees: "she didn't like pretence of any kind. She was never irritable, never impatient. She had time for me: she had time for everyone she met. She was absolutely sweet."

The year after Epstein's death, Kathleen befriended a handsome young American poet, Samuel Menashe, who was in London for the publication of his first book. Menashe remembers her vividly. "She was tall and she wore her hair straight down like a young girl and she had a slender figure. It was a strange contrast. She was 59 then, with this old woman's face, but from the back she was like a ballerina, six feet tall. She was very hospitable and she loved parties. She loved elegant living. She had beautiful hands. If she peeled an apple . . . everything was done with an aesthetic sense. Nothing was brash." Her association with Menashe was teasing, somewhat flirtatious. She would offer him lavish gifts, a Burberry coat, a birthday party, an Epstein drawing, which never materialized; "she had the thrill of her own generosity, then, just by saying what she was going to give." In return, he wrote poems in her honor and took her advice about his work.

Menashe remembers that she went every Sunday to visit

Theodore's and Esther's graves and that she told him that being in between her husband and her son had been "like being ground by a millstone." He recalls her saying that "the father had had no joy in the son." Nevertheless, Menashe's memory is of someone who loved life. "I love a person who enjoys what he has, and she did. She enjoyed life. The teasing was fine with me. It was wonderful to have her as a friend, but because of the sparring I wouldn't have liked to be a child of hers."

Kathleen continued to support Beth Lipkin and the young Roland Joffe, as well as helping out her sister Helen whenever she could. When Kathleen was away for Beth's birthday, she left presents of banknotes and scent and gloves hidden in bags in a cupboard, then wrote to tell her where to find them. For one of Kitty's birthdays she "scoured the shops for forget-me-nots and rosebuds and packed a charming present of Caron perfume and Dior stockings and a rosebud scarf and handkerchief and two £5 notes packed in the loose front of an old Victorian mirror painted with daisies and forget-me-nots." She was "immensely generous—to a fault," says Kitty. "When I was first married to Wynne we were hard up and Kathleen found an old Treasury briefcase and she stuffed it with £10 notes and we just used to dip into it." She sent parcels to her brothers and sisters: a red jersey to Sylvia, a record of Bartók, and a book of Constable sketches to Mavin. To Douglas she sent money tucked between the pages of a new biography of Napoleon. "You know your courage must be an inspiration to a lot of people besides me," he wrote to her in gratitude.

She took picnic lunches to an elderly friend, with specially cooked chicken in herbs and wine, pâté, salad, fresh strawberries, and strawberry mousse (it amused and exasperated her that the "chicken was pushed to one side with a truly terrible gesture as being too much like 'meals on wheels' "). Dr. Valerie Cowie remembers countless picnics. "Once we went to Richmond Park and a park keeper came up and said: 'You can't eat that here: it's no picnicking,' and Kathleen held out a glass of wine to him and said: 'Do join us, officer.' But it didn't work and he got very firm, which surprised her. She was always offering blandishments to people."

Epstein had been an avid collector throughout his life, often unable to resist adding to his hoard of tribal and ancient sculpture even at times when both Mrs. Epstein and Kathleen were struggling to maintain their households. In accordance with his will, this collection was sold after his death. Now Kathleen herself began to collect, but she did not do so alone. Sally Ryan was an American sculptor who came to England in 1935, when she met Epstein. She was the granddaughter of Thomas Fortune Ryan, an Irish-American entrepreneur with interests in tobacco, rubber, iron, gold, railroads, and shipping. A taste for sculpture ran in the family: Fortune Ryan had commissioned a portrait bust of himself by Rodin, which is now in the Tate Gallery collection in London. His lawyer and friend, John Quinn, had been one of Epstein's earliest and most important collectors. Sally's maternal uncle was the painter Augustus Vincent Tack, much of whose work is in the Phillips Collection in Washington, D.C. Tack was instrumental in the careers of Jackson Pollock, Mark Rothko, and others. Another uncle was the famously flamboyant John Barry Ryan, who, in the early 1930s, inherited some thirty million dollars and is said to have given Rolls-Royce cars to many of his friends.

Sally Ryan was a very shy, retiring person; a dedicated and talented sculptor who dressed modestly and kept out of the limelight. She was sandy-haired, blue-eyed, fair-skinned, and slender. Her parents had had a highly publicized divorce in the 1920s, involving George Maxwell, president of the American Society of Composers and Performers, who had started life as a piano tuner in Glasgow. He was alleged to have sent scurrilous love letters to Mrs. Ryan; the case caused a sensation in New York society, with Ryan himself being accused of forging the letters, although all the evidence pointed to Maxwell. Sally's mother had fled to get away from the publicity. She and her husband were divorced, leaving her in a state of nervous collapse with the care of six children. Having seen what publicity had done to her mother, Sally Ryan was always determined to avoid it herself.

Toward the end of the 1950s, Sally Ryan learned that she had an incurable cancer of the throat and decided that she wished to leave an art collection as part of her legacy. In England, where she spent most of her time, she and Kathleen worked informally together, visiting dealers and auctions. By the mid-1960s, Kathleen had begun to refer to their purchases as the Garman Ryan Collection. Curiously, no letters from Sally Ryan to Kathleen survive (Beth Lipkin took charge of Kathleen's papers upon her death, and those letters and documents which remain were selected for preservation by her. They consist mainly of letters to herself from Kathleen written during the 1960s and '70s). Only one, brief note remains to attest to the Garman Ryan joint enterprise. Having bought a Bonnard riverscape, Sally Ryan enclosed a slip: "Dearest Kathy, This BONNARD landscape 'La Seine à Vernon' is for you, with my love—Sally. Easter 1966." At her death, in 1968, she bequeathed all her works of art to Kathleen, as well as fifty thousand dollars in cash. (In America she had lived with a woman, Ellen Bullock, to whom she bequeathed a life interest in a property in New York.)

To what extent Kathleen and Sally bought together or sought approval for separate finds is unknown. The two women motored to country salerooms in a car with a driver. They evidently enjoyed themselves, as their friend John Lade recalls, "The telephone rang one evening and it was Kathleen. She was with Sally Ryan and she wanted to bring her to see Theo's pictures [Lade owned several]. An American car with a driver waited below. It was all great fun, we all went on the the Dorchester, where Richard Burton was coming out of the elevator. He embraced Kathleen. Then we went up to Sally's suite where there were all these pictures; the pictures which are now in Walsall, or some of them."

Some of Kathleen's things also ended up in the collection: a Modigliani drawing that the artist had given to Epstein in Paris in 1912, two Matthew Smiths, numerous paintings by Theo. Works by Lucian Freud, Epstein, and by Lorna's son Michael Wishart also found their way into the canon. "The Garman Ryan Collection was basically the stuff from the dining room at Hyde Park Gate," recalls Jackie Epstein. "What was our fruit bowl is now in a glass case, with a label." Kitty, too, recognizes in the display several pieces familiar from 272 King's Road days. Kathleen had sold Hyde Park Gate in 1965, moving to Elm

Park Gardens in Chelsea. There was less room, here, for the paintings and ethnographic pieces that remained after the dispersal of the bulk of Epstein's collection, and room had to be found for the pictures she and Sally Ryan had collected, but Kathleen crammed things in. To Beth Lipkin she wrote that some visitors "were bemused and enchanted by the sculpture and paintings and exclaimed 'this room should be kept forever just like this for a glorious museum.' "

When she decided that the collection should be accessible to the public, Kathleen's first thought was to house the Garman Ryan Collection at Oakeswell Hall. But a visit showed that this could not be, for the house had been demolished, leaving behind only its tall stone gateposts. Yet the idea of creating a home for the collection in her native Black Country had taken hold, and in 1972 she was shown the first floor of Walsall Library. This long room, with its carved and vaulted ceiling, was to house the collection until its move to the new premises in 1999. The artworks included in Sally Ryan's will—their value was put at three hundred thousand dollars—together with the pieces already in Kathleen's possession, were to be put on display in Walsall. "I would like it to be a model of its kind at the same time striking a note of intimacy and spontaneity that will appeal to all ages," Kathleen wrote. "I feel we are dealing with dreams and are about to house them in a solid Midlands setting for posterity. How delightful."

In addition to the Garman Ryan Collection, Kathleen was occupied with the Little Gallery, a commercial venture that she took on in 1969. The premises were in Kensington Church

*Kathleen with her former son-in-law, Lucian Freud,
during the early 1970s, when she donated the
Garman Ryan Collection to Walsall.*

Walk, and the gallery specialized in group shows with imaginative themes and titles: An Anthology of Trees was a show of
drawings and watercolors from the eighteenth to the twentieth
century; Of Music and Musicians followed. Kathleen especially
enjoyed assembling portraits for a show dedicated to pictures of
writers. "I can hardly wait to get on to the literary exhibition,"
she wrote to Beth, "Baudelaire, Verlaine, Thackeray, E. M.
Forster, with letters and illustrations..." Each year the Little
Gallery put on a special Christmas exhibition. The Garman
Ryan Collection is also arranged thematically, and it seems that
at least some of the stock from the Little Gallery found its way
into the permanent Garman Ryan Collection. Beth Lipkin was

installed in the gallery, and it may be that the enterprise was thought up to give her something to do. But Kathleen was clearly in charge, overall. "Please pay ardent attention to the affairs of The Little Gallery," she wrote to Beth from Italy, "and have the mauve water colour mount changed *at once*."

Kathleen had been one of the last bastions of the old, bohemian Chelsea. What had been a shabby enclave of artists' studios and the rented rooms of Oakley Street was already becoming chic and expensive, obliging her and Beth to move farther from the center of London. In 1973, they moved to Redgrave Road in Putney, to a house on the corner of a quiet street of brick villas. Here Kathleen planted climbing roses and clematis and her favorite lilies. She decorated the house. "For me dust and dirt spell decay and slow dissolution and until we fall into the dust ourselves we must make our surroundings as beautiful as we can," she wrote. In London, she saw friends, visited galleries, listened to music. Dr. Valerie Cowie, whose husband had died in 1970, lived only around the corner. Sometimes Kathleen went to church at Holy Trinity, Sloane Street, where Theo had been a choirboy. The Basil Street Hotel was a favorite meeting place for cocktails in the evening, sometimes before a concert, opera, or ballet. Poetry remained a passion, and she dined with T. S. Eliot and lunched with Allen Tate, when he came over from America (she still called the midday meal "luncheon"). Another frequent companion was Richard Buckle, who worked on his Epstein book in close cooperation with Kathleen, and there were various representatives from museums and galleries, her nieces and nephews and old friends such as John Lade and his wife, Susan, who had been so kind to Theo.

Kathleen spent time at her house in Torri del Benaco and in France. She sent Beth a postcard from Paris: "I put four roses on Van Gogh's grave and was glad there was such a lovely sweep of open country and sky all around and thought of Harting. I did not know we were going to see the poor little humble attic where he died and it was too much for me altogether and couldn't sleep all night for crying. It was all so familiar but it seemed wrong really that everyone can tramp in there into that place of suffering even though we go with gratitude and homage." Such an admission of intense feeling was rare; mostly she was unruffled, or seemed so.

"Such colours glowing, smouldering, sparkling in the autumn sunlight as I have never seen," she wrote from Lake Garda, one November. "I drink it all in like wine burning through my veins and then feel melancholy that I cannot share it with the ones I love." She was particularly happy when her granddaughters came to stay, or Roland Joffe and his family; she was delighted when he married the actress Jane Lapotaire ("Jane stole the whole show... top marks to Jane for real, good, moving acting. Really splendid," said Kathleen, after watching her on television). Joffe remembers her walking everywhere, striding eight kilometers up a hill to see a little church. "Kathleen thought every stone should be picked up and looked under, because there'd be an adventure underneath," he recalls. Even waking was a pleasure to her. "The birds and I have woken up at the same moment and the last of the big white magnolias are unfolding on the top of the tree on a level with the bedroom window. Soon the trees will be overflowing into the house. The wistaria already twines on its way in through the shutters in the music room," she wrote during her last summer in Italy.

When she was with Helen, whose sight was failing, Kathleen read aloud to her. "We start each day with the same procedure, of my choosing i.e. Pascal's Pensées, in French for half an hour interspersed with lunch, discussion as to the exact meaning of his thoughts; nearly an hour of John Donne's poems, devotions and sermons which are thrilling to read aloud; and then when I make the tea we get down to Mrs Gaskell and bask in her gentle, sunny, straightforward story after the heights and depths of turbulent logic of the others." No cocoa, crosswords, or crochet for Kathleen's old age. She did not compromise her intellectual standards, nor her aesthetic ones. The house in Putney, like that in Italy, was furnished simply but with great style: her bedroom, where she liked to read, was hung with pale blue watered silk. She still wore her hair as she always had, naturally straight and long in an era when most of her contemporaries were slaves to the stiff coiffure of the shampoo and set. A young woman who met her in the 1970s was very struck by her appearance. "A very long neck, huge eyes and the hair cut straight: she was startling and very impressive."

"Kathleen wasn't all charm," says Roland Joffe. "People who are all charm don't stay in the mind, but combine that charm with a certain element of steel: that really works. That was what she was like." Employing that steel, she faced her own death from cancer with characteristic equanimity and courage. Valerie Cowie remembers Kathleen calmly taking her hand and holding it to her side, where she felt a tumor "the size of a grapefruit." When Kathleen thought it was time, she summoned Roland Joffe to her bedroom. "She said she was going to die and she didn't want to go to hospital. She said she knew she could rely on me to sort out certain things for her and she told me who would come and claim what, and in what order they'd arrive, once she was gone. All of which proved correct. And then she reached into the bedside cabinet and pulled out a bottle of champagne."

Kathleen died in August 1979 and was buried at Harting, with Theo and Esther.*

---

* A grave that her sister Helen and Beth Lipkin both expressed a wish to share, so as to be laid to rest with their beloved Kathleen.

Chapter Fourteen

NOTHING IN MODERATION

D ouglas Garman remained true to the cause of communism till the end of his life. As educational officer for
the Party during the late 1940s, he had lectured at trade union
branches and literary institutes up and down the country and in
London at the highly regarded Left Book Club. He also set up a
number of residential schools, teaching the fundamentals of
Marxism to hundreds of industrial workers from all parts of the
country. His letters to Paddy from the period are full of enthusiasm for the students and the cause. "Teaching is really my
work. I wish you and I were running a plan like this as a permanent Party school...the fact of teaching, of being able to get
over to other comrades what I know to be exciting and important, stimulates me enormously ..." He also enjoyed walking and
swimming (which he called bathing, in true Edwardian fashion)
with miners from Lanarkshire and Welsh steel workers. The
schools rejuvenated him. "The fact is, I don't want to get old," he
told Paddy. "Keep loving me as much as I do you and we'll be in-

vulnerable. And don't be lonely. Believe me, this work does help to bring about what we both try to live for." Colleagues remembered him as an inspiring teacher, with a gift of lucid exposition and a most unusual power being able to enter other people's minds and draw out their doubts and difficulties. He was ready at any time to talk with individual students. When comrades speak of Douglas, two words come up: kindness and integrity.

In 1950, though, things went wrong. Garman had been finding himself increasingly at odds with the bureaucratic Party leadership of Harry Pollitt and Emile Burns, which made it more and more difficult for him to remain education officer. An entry in a notebook explains his change of direction; "in 1950 I decided that, if I could now write, in such a way as to give expression to the far greater understanding of the class struggle and my much deepened conviction of its necessity, my participation thro' writing would be more effective." With this objective in mind, he and Paddy now moved out of London, first to Sussex and then to Dorset. He made copious notes and drafts for a semiautobiographical novel which dealt with political issues through the changing beliefs of a large family and also worked on a polemic, provisionally entitled "The Necessity of Revolution." "My intellectual conviction of the truth of Marxism is, within the limits of my capacity for knowing and thinking, absolute," he noted in October 1951. He also read a great deal of Keats, Tolstoy, Coleridge, and the Baudelaire Kathleen sent him. Like Mary, he listened to Mozart.

But neither of his works was to see publication, for his troubles went deeper than those revolving around Party bureaucracy. In one of the letters he wrote to Paddy, Douglas asked her to telephone their doctor for a prescription of fifty Benzedrine, an amphetamine. He made several mentions of his need for solitude, referring to the enchantment of being alone and the strain of being obliged to be extroverted. Listlessness extreme enough to require medication and a strong desire to be alone are among the symptoms of clinical depression. The crisis which coincided with his move to Dorset was not his first experience of the illness, but this was a more serious breakdown than he had had before. "I cannot sing, for my throat is hoarse with slogans," he had written in one of his poems; the words seem prophetic. He was left much incapacitated, calling upon Paddy's considerable reserves of efficiency and good sense.

Disheartened by what he perceived as the inadequacy of his writing, he turned to farming. But Douglas had no experience of livestock husbandry, and he wore himself out in the effort. The Garman sense of humor did not desert him, even in these strained circumstances; he had seven pigs and called them each by name: Mary, Kathleen, Sylvia, Ros, Helen, Ruth, and Lorna. He also started one of the few agricultural branches of the Communist Party. "Lorna said he used to strut about the country with his silver topped cane," remembers one member of the family, "he was quite pompous."

Toward the end of his life he lived very modestly, writing the text for some of the renowned Shell Guides to England, as well as making a translation of Flaubert's *Sentimental Education*. Old friends came to see him and were warmly received. Among them were Alick West and his wife, Liz. Douglas had met West at a political meeting at the Conway Hall in the early 1930s, and they corresponded for many years. West had been lodging with Eric Hobsbawm in London, prior to marrying the much younger Liz.

Meeting Douglas for the first time, Liz felt he was very much like an English country gentleman of the eighteenth century, with his perfect manners and wide knowledge. "He had the effect of making you want to live ... I think this is why he was an exceptional person," she said. Liz West also felt that Douglas's truthfulness "must have isolated him, because there were few people who could live up to it."

Douglas died at sixty-six. At the funeral in December 1969, Alick West said a few words:

> He loved life with a humour which ranged from the exuberant to the sardonic, and with an intelligence which knew its heights and depths and faced them with courage. He strove to live with all his consuming energy and to make others live. He could not endure that anyone should exist in indifference. Where he was, he quickened the life around him into pleasure and gaiety, laughter and wit, and with honesty that went to the very heart. In his friendship was the unsparing generosity of truth. Poetry inspired him. He made himself a Marxist because in Marxism and revolution he saw the same promise of life as in poetry. In the Party schools which he created he enabled others to see it also. There are men and women throughout the land who will never forget him.

Paddy survived him by over thirty years.

$\mathcal{A}$s to the other siblings, Mavin Garman had the dash and good looks of a film star from the golden age of Hollywood. While he was still in his teens, in the mid-1920s, he sailed for Brazil to become a cowboy. Before long, he was running a twelve-thousand-acre estate, and married. He brought his bride back to England, and they had a daughter, named Lorna Ruth. The marriage ended. Like Douglas, he joined the Communist Party, meeting his life companion, Rosa Slater, through their shared interest in politics.

After running a large farm by the side of the sea in Dorset, he and Rosa opened a series of shops selling antiques and curios; like his sisters, Mavin had an excellent eye and juxtaposed ancient and modern, local and exotic pieces with simple elegance. He relished what he called "the art of mere living," sweeping others up in that delight, whether it was in laughter and the company of friends, a favorite book, or a winter picnic in the woods, reached on horseback. Mavin and Rosa's son Sebastian was the only child born of the nine Garmans who lived to carry on the family name.[*]

Rosalind Garman preferred to live her life out of the spotlight. As a young woman she worked for Sir Phillip Hendy, who was to become head of the National Gallery, before enjoying considerable success as a dressmaker. She married a very good-looking Italian Scot called Paolo Brown, whose mother, Agnese, had been a figure in the Florentine art world, numbering Giorgio de Chirico among her dearest friends. For many years Rosalind and Paolo ran a garage in Surrey, where they stored some of Epstein and Kathleen's paintings for safety during the Second World War. Later, Rosalind lived in Dorset as a widow for some thirty years, where she baked all her own bread and created a glo-

---

[*] Rosa changed her name to Garman. Sebastian's son, Theodore, brings the Garman name into the twenty-first century.

rious walled rose garden, corresponding about varieties with the well-known rose-grower Hilda Murrel. Rosalind is remembered with tremendous affection for her practicality and great kindness; she was the favorite aunt of numbers of the Garmans' children. Her daughter believes she was "the sanest of the lot."

Ruth Garman stayed on in remote Herefordshire after Mrs. Garman moved to Sussex. The gentlest of the sisters, she was also the most wild: "If *only* Ruthie could go into a pub without getting pregnant!" Lorna is said to have sighed. Ruth had five children, only two of them born during her marriage. The eldest son was brought up to believe that his father had been an Admiral Reed; only when he came of age did his mother admit that there had been no such person and she had simply taken the name from an inn she passed as she came out of the nursing home where she'd had the baby.

When they were young, Ruth and Lorna would meet in London to go dancing, and it was here that Ruth had a fling with Augustus John and another with the Caribbean pianist Leslie Hutchinson, known as Hutch. Hutch had been the protégé and lover of Cole Porter, and went on to have affairs with Ivor Novello, Edwina Mountbatten, Tallulah Bankhead, and Merle Oberon. One of Ruth's children was allegedly Hutch's child, although a tango teacher from Hereford may have been the father. A daughter, Stephanie, died a suicide at twenty-two; a drawing of her by her cousin Theo is in the Garman Ryan Collection. Lorna helped her sister out, but Ruth was never well-off. She lived in a cottage without electricity or hot water overlooking a cherry orchard at Blakemore, near Hereford.

"Low pubs and high characters: that was her thing," recalls her son, William. "But she was immensely kind and there would be no more constant visitor than Mum to anyone old or ill."

Her niece Kitty remembers her aunt Sylvia as the loneliest and perhaps the strangest of them all. Graceful and very tall, people would turn in the streets to look at her, even into her seventies. While driving an ambulance in London during the closing months of the First World War, she met a fellow female volunteer, Jo Cubitt, who was to become her lifelong companion. They lived in isolation in a primitive cottage in the woods near Wareham in Dorset, where they kept bees and drew water from a well. They adopted a boy, the son of a couple who kept a local shop, but the two women separated for a time when Sylvia married a sailor who had appeared at the door, lost, to ask directions. The marriage did not last, and Jo was reinstated. Kitty recalls an occasion when a gown that Sylvia had ordered arrived in the mail. It was an evening dress in dark blue velvet, low-backed and with a cowl neck. Sylvia tried the dress on, then burst into tears and ran out into the forest still wearing the gown, weeping, because no one ever invited her out.

Sylvia drove an open-topped Sunbeam Talbot car, wearing an aviator's hat with flaps while she was at the wheel, always with a supply of Sporting & Military dark chocolate in the glove compartment. A family rumor persists that she was a girlfriend— possibly the only girlfriend—of T. E. Lawrence. Visitors recall his portrait, by Augustus John, hanging in the house. Lorna's son Michael described his aunt as "a lesbian with a moustache, the image of George Sand, who with or without the lash managed to seduce her idol T. E. Lawrence once only..." It is said that he bequeathed to her the original drawings from his book *The Seven Pillars of Wisdom,* together with his collection of pipes. Some believe that Lawrence was on his way to take tea with her when he was killed in a motorbike collision. But Jeremy Wilson, the acknowledged expert on Lawrence, has never heard of Sylvia Garman. As to the pipes legacy, Wilson adds that T. E. Lawrence "did not smoke a pipe at any time in his life, and never showed any interest (that is recorded) in smoking pipes."

Lorna's grandchildren refer to the latter half of her life as the time "after her naughties." Her conversion to the Roman Catholic church took deep root following the end of the affair with Lucian Freud. "She became very, very Catholic and gave away all her jewellry and went to mass every day and renounced all her lovers," says Kathleen's daughter Kitty. "She gave me a lovely dark amethyst bracelet. She was very extravagant in all her gestures, like my mother." Lorna's faith had its roots in the late 1930s, when Mary had come to stay with the Wisharts, full of the fervor of Spanish worship. The two sisters can have spent very little time together, since Mary was thirteen years the senior and had lived abroad for most of Lorna's adult life, but there was a close bond between them. They both embraced the faith with the same gleeful piety that they had brought to playing at church during their childhood at Oakeswell, a game of make-believe in which Mary always officiated. Lorna was to remain the one person in her family for whom Mary had any affection.

From the end of the 1940s, Lorna immersed herself in God. She took up Llewelyn Powys's advice and turned also to gardening. After early mass in the mornings she went straight out to the garden, where she planted great clumps of Japanese anemones, foxgloves, hibiscus; honeysuckle and roses which spilled their petals onto the grass; tangles of bright orange poppies, nigella, bluebells. The garden at Marsh Farm, Binsted, was divided into a series of outdoor rooms, none of them overlooked by another. There was a space walled with Sussex flint and old brick, as well as a shady, sunken enclave where flagged paths led through ancient yews. There was a little orchard where Wishart kept bees. Classical statues were dotted along paths edged with box. Occasional gates were set into walls and hedges, through which to glimpse the distant prospect of meadows. A shell pink *Clematis montana* twined its way over a door painted the color of verdigris. There was a wooden summerhouse and an old air-raid shelter almost concealed by overgrowing plants, where Lorna used to hide things and, occasionally, herself. At the back of the house a long lawn with a ha-ha gave onto a grazing paddock.

When she wasn't out of doors, Lorna was usually to be found in the kitchen. She pioneered of a style of farmhouse kitchen which has become nearly ubiquitous in Britain today. Marsh Farm is a stone-flagged, Elizabethan house, and the kitchen with its Aga (a huge range-style enamel stove) was always the warmest room, in itself a compelling reason to sit there. If there was a front door or a front hall at Marsh Farm, no one ever used them. Visitors came through a tiny courtyard, by clumps of bracken and big stones brought back from the beach, to the scullery door and thence straight into the kitchen. Unlike her

Lorna in her garden in the late 1940s.

peers, who would receive visitors in the drawing room, Lorna would generally be sitting in her chair to the right of the stove, a cigarette in her hand. At the time, kitchens were intended to be utilitarian rather than pleasing to the eye. Influenced by her visits to the Mediterranean, Lorna put pots of geraniums on the windowsills, lemons in a wooden bowl, and a riot of mismatched china on the bright blue painted dresser. There were ceramic figures of cats and a stuffed owl on the mantelpiece, which looked down at a stuffed cobra. Blue- and white-striped Cornishware dangled from hooks. Behind the Aga were tiles that didn't match, some of them hand-painted by Lorna, as was the Latin inscription over the door: *Divinum auxilium maneat semper nobiscum* ("Divine help remain always with us"). The brickwork was painted the color of Cornish cream.

Over time, the rest of Marsh Farm grew dustier and dustier. "It was like a house from the 1920s," recalls one visitor. "I don't think anyone ever cleaned it at all." Wishart amassed huge collections of matchsticks and newspapers and parcels of books, some of which remained unopened in the attics. "It was quite cobwebby and Wish would never throw anything away," says another friend. "It was like Miss Haversham's."

Lorna was as unconventional as her house. She would go out riding for the morning, then decide on a whim to drive to London, heading off at speed to have lunch at the Café Royal or Chez Victor, still dressed in her long riding boots and tweed hacking jacket. There would be picnics in all seasons, always with Wishart's favorite raspberry jam sandwiches. Or Lorna would take it into her head to go out riding in the dead of night, muffling her horse's hooves with dusters so that people in the village wouldn't hear her go by. She never, never stood in a queue: she simply went straight to the front and no one ever seemed to challenge her. She liked Elvis, the English royal family, nuns, and later, Margaret Thatcher.

Lorna took people in, especially troubled youngsters, and supported them utterly, but she could be cruel, too. One eight-year-old, the daughter of Lorna's best friend, was swimming in the sea with her, when Lorna held her under the water for so long that she gasped and choked. "Lorna tried to drown me, when I was a child . . . she said I was a spoilt little brat, and she thought I deserved a ducking. She was probably right," the girl remembers. She nevertheless grew up to regard Lorna as an inspiring friend and close companion.

Douglas Garman's stepdaughter Rona was one of many younger members of the extended family who spent a lot of time at Marsh Farm, along with Ruth's son William, Mary's two girls, and Kitty. Some of these young people followed Lorna into the Catholic faith, and one of them even took holy orders. Lorna remained close to Mavin and Ruth, her favorite siblings, and to her sister Rosalind, with whom she competed as to whose country garden was the more beautiful. (Lorna and Kathleen never got on, according to Rona.) "She has been presented as a Femme Fatale, but that misses out parts of her," says Kitty. "She was jolly, too, and there was an inconsequential, flippant side; also she was maternal and loved to be with children and adolescents particularly, giving them treats; but always with a little acidity in it. That was what Lucian meant when he said of her that she was the first person he had met who he could care for *and* respect. She had this effect on many other of her lovers and knew all their draw-backs."

If people knew her failings, they generally weren't put off by them. She was magnetic, a force field, but her passionate nature could make her jealous and sharp. "I hardly ever remember her saying she was sorry about anything and when she shot it was with both barrels," recalls a friend. "I don't think many people confronted her, ever. They wouldn't have dared." One bohemian young woman who was about to be married was advised by Lorna to cut off her long hair before the wedding and to dress for the registry office in a very conventional tweed suit. It did not occur to her not to follow the older woman's advice, although she was miserable at the effect, which was quite out of character and unbecoming.* Lorna was a guest at the wedding, and all eyes followed her instead of the bride, as she had doubtless intended. To the few who didn't come under her spell, she seemed cold, manipulative. Her gift for intuition could be perceived as witchlike.

Lorna surrounded herself with a menagerie. At one stage, there were some half-tame owls in an outhouse, and she made a pet of a wild fox. She always had very difficult horses and bad dogs, three or four at a time: greyhounds, Alsatians, Chihuahuas, Jack Russell terriers, a poodle called Sappho, all very unruly. A favorite horse was partial to a pint of Guinness, a taste which Lorna indulged at the village pub. Riding became a passion: Lorna advised the young woman who had supplanted her in Lucian Freud's attentions to "Get a horse and become a Catholic. You'll never regret either." She never bothered with a hard hat, as if the element of danger was part of what she enjoyed. "She always looked as though she was about to jump on her charger and gallop off into the mist," recalls the painter Anne Dunn.

Animals appeared in many of her pictures, too. Upstairs in the attics was where Lorna painted. There were canvases stacked around the walls, paraffin stoves, old mattresses, pictures by Lucian Freud, and first editions of books bought by Wishart, never unwrapped from their brown paper. In the farthest room, she set up an altar and an easel. She painted serpents and unicorns, white horses, owls, squirrels, and the cat which Lucian Freud had given her. Her pictures were like illustrations to fairy tales: highly romantic, mystical. Death and the

---

* An incident which recalls F. Scott Fitzgerald's short story "Bernice Bobs Her Hair."

maiden was a favorite theme. She painted Leda and the swan; Mary, Queen of Scots; saints and sacred hearts. There was a picture of Arundel by moonlight, like a many-towered Camelot, the silvery river winding away into the distance. One especially striking painting shows the back view of a bride running into the night, scattering her bouquet, while a tiger snarls in the foreground. Lorna had great confidence in her own gifts. "She genuinely did think she was as talented as Lucian, till the end of her life," remembers one friend who knew them both. "She certainly thought she was a better artist than her son Michael."

Nevertheless, she painted only sporadically, preferring to be out of doors walking or on her horse or in her garden. Whatever the day held, the first thing she did in the morning was put on mascara and eyeliner, even into her eighties. She liked lipstick. She was careless of her appearance and yet vain at the same time. There were nods to fashion: from the 1970s she knotted Indian silk scarves around her throat or wore sleeveless suede jerkins with heavy fringes. (She liked these long waistcoats so much that she bought two, one in fawn, the other in lilac.) When she went to take a grandson out from prep school in the 1960s, everyone stared at the dramatic woman with the long dark hair and eyeliner. Afterward the other little boys asked him: "Who was that beatnik?" Mostly she wore trousers and knee boots; she tucked a six-inch hat pin into her boot whenever she went riding, in case she needed to defend herself.

Lorna and her husband quarreled over Catholicism and socialism, although he relinquished his faith in Marxism after the Hungarian uprising of 1956 and she became less fervent with age. At one time, she became close to the Poor Clares at Arundel, an order of nuns who live very austerely, sleeping on only thin straw mattresses to cushion them against their stone beds. Learning of this, Lorna asked her husband to send a nice big load of straw up to the convent for them; he sent a whole trailerful, but of barley straw, which causes insufferable itching. Lorna enjoyed repeating this story, amid much laughter. "She had a terribly wicked sense of humor," remembers a friend.

Lorna was always the leader, the center of gravity. "She had this commanding presence, she took over," says Anne Dunn. "She was full of physical courage. I don't think she knew what it was not to be. She never retreated." If she suddenly wanted to go riding at midnight in midsummer, then there was always someone willing to go along with her. "Nothing in moderation!" she'd call, galloping up the hills, dark hair flying. Llewelyn Powys had once written to her, telling her off for drinking too much. "I hope you are not vexed with me ... I was born a stern moralist and I could never abide the idea of beautiful girls obscuring their eager senses with wine—I like them to drink till their eyes shine but never, never to dull those little cries of anguish and ecstacy that make the sweetest music ... do not be careless of your life." But careless she was, and she did drink.

All her life she swam in high, cold waves, oblivious to danger and to what others thought of her. She didn't know what embarrassment was. "She had a tremendous pleasure in life: galloping on her horse, her garden, her lovers, driving fast cars," according to Anne Dunn. "She enjoyed her existence. She loved life." But her daughter, Yasmin, offers a different picture. Lorna had been painfully shy during the early years of her marriage, and it was a trait she never fully lost. Toward the end of her life, she became almost reclusive. Yasmin recalls how Lorna would entertain guests to tea and make everything perfect for them: the thinnest cucumber sandwiches and Lorna at her most funny and compelling and marvelous. Everyone would be charmed. But after they had gone, she would sag with exhaustion from the effort of seeing people. "She wasn't an actress," says Yasmin revealingly, "but she did act."

During the last years of her marriage, she and Wishart discovered a new closeness, and she even went so far as to tell one friend that it like a second honeymoon. Rather to everyone's surprise, she became a devoted and efficient nurse to her husband during his final illness in 1987. Her son Michael died from cancer in 1996, and at the funeral in Arundel Cathedral people turned to stare as Lorna came in. She looked wild with grief but still beautiful, still impetuous and proud. Laurie Lee, who was to die the following year, wrote to commiserate, knowing how stricken she would be.* Not long after Michael's death, she was involved in a serious car accident, from which she never fully recovered.

Lorna had been the youngest of the siblings, and she was the last to die, on the day after her eighty-ninth birthday, in January 2000. Her ashes were taken to Cornwall, where she had always been so happy. A wooden statue of the Virgin Mary, carved by her, still stands overlooking the pond in the woods at Binsted. Closing the circle, Lorna's family donated a piece of her sculpture to the Garman Ryan Collection in Walsall, near Oakeswell Hall, where it all began.

---

* Lee remained married to Helen's daughter, Kathy, throughout his life. In one of the coincidences that seem to characterize the Garmans' lives, their daughter was born on the very same day as Yasmin's baby girl, Lee's grandchild with Lorna.

APPENDIX
THEO AND ESTHER

Kathleen was too dignified and too courageous to mourn the deaths of her two children publicly, although she confided some of her agony in private notebooks, copying out favorite lines by Thomas Hardy, as well as poetry of her own. She did not talk about either tragedy. How Theo died is something the family still finds too painful to discuss.

In *A Shared Vision*, the catalogue to the Garman Ryan Collection at the New Art Gallery in Walsall, there is a reference to Theo's having attempted to take his own life, following a prolonged period of depression in 1945. "Despite some recognition as a professional artist over the next few years," the text continues, "including a successful one-person exhibition at the Redfern Gallery in 1950 (for which Matthew Smith wrote the catalogue introduction), Theo's mental health declined. He died of a heart attack while being admitted to hospital in 1954, in circumstances that have never been fully clarified."

It is now half a century since Theo died, and only a handful of people remember what happened. Some of the people I spoke to about Kathleen and her brothers and sisters brought up that tragic event. Everyone spoke warmly of Theo, and it says a great deal about the uniqueness of his nature and talents that he is still remembered with such feeling.

In the course of researching this book, I visited South Harting, the village where Mrs. Garman and Miss Thomas had lived. Surrounded by woods and hills, it retains an old-fashioned charm, with its characteristic Sussex flint walls and tile-hung cottages and warm brickwork. Their house, Vine Cottage, stands at the foot of the hamlet, near the pub; the church is up the hill, at the southern edge of Harting. I'd hoped to encounter some residents who might still remember Mrs. Garman and her children, who had been such frequent visitors, but the place seemed quite deserted. The only sound was the songbirds calling in the empty graveyard where Kathleen, Esther, and Theo are buried. But as I strolled down to the village from the church, I noticed an elderly man cross the road and go into the telephone box. I waited until he'd finished making his call and then approached him.

Oh yes, he said, he knew the Garmans. I asked a few questions about them all, but it became evident that it was Theo he wanted to talk about, and he was not to be deflected. "Theo was an untidy chap," the man said. "Lived with the Frasers at Rosemary Cottage. It's called Hill House now. They got fed up with him. Then he lived in a caravan, see, but his mother kept on at him that he couldn't live like that. 'I'm not daft,' he said, 'but they'll drive you daft.' All he wanted to do was paint. They killed him, didn't they. Drugged him."

According to his death certificate, Theodore Jacob Garman, aged twenty-nine years, of 272 King's Road, Chelsea, died of edema of the lungs from acute heart failure during a maniacal attack while suffering from schizophrenia. There was an inquest into the death, details of which were reported in newspapers of the time. The various reports make harrowing reading. The London *Evening News* for February 1, 1954, carried an account of the inquest on its front page:

'SPLIT MIND' ARTIST IN AMBULANCE STRUGGLE

The West London Coroner, Mr H Neville Stafford, said at Hammersmith today that an inquest on 29-year old Theodore Jacob Garman, of King's Road, Chelsea, artist protégé of Epstein, was being held because an anonymous letter had been received at Chelsea police station. Part of the letter, read to the jury, said: "The numerous friends and acquaintances, in Chelsea at least, of Theodore Garman are horrified and appalled by the barbarous manner in which he was virtually hounded to death." Garman died in St Stephen's Hospital, Chelsea, where he had been taken when he became unconscious after a doctor had given him a sedative to quiet him. Inspector Graham Goddard said he was satisfied that there was no truth in allegations in the anonymous letter about violence. The jury returned a verdict of death from natural causes.

'Drug His Food' Request to Mother

Dr A D O'Connor, of Coleherne Court, South Kensington, said Garman, powerfully built, over 6ft tall and weighing about 16st., was a schizophrenic (split personality). He had been twice a voluntary patient in a mental hospital. "On January 19," said Dr O'Connor, "Mrs Garman, his mother, asked me to see him. I tried to persuade him to go as a voluntary patient to a private mental hospital." Continuing, Dr O'Connor said he failed to persuade Garman to sign a voluntary order so his mother signed an emergency order. He attempted to get Garman to the hospital by means of a "ruse." On advice from a doctor at the hospital he asked the mother to drug Garman's food. Then two male nurses called at the house with an ambulance to take him away.

A Struggle Beside Ambulance

"I was watching from my car," the doctor added, "and I saw Garman leave his home with the two nurses. At first he was walking quietly, but a struggle developed when they got to the ambulance. A crowd gathered. I did not see the people interfere but I have heard since that they did. Some of them were known to Garman." Garman got back into the house. There, said the doctor, he and police officers tried to persuade Garman to get into the ambulance. He gave him two injections to restrain his violence. After the second Garman became unconscious and he was taken to St Stephen's Hospital. Dr Skene Keith, pathologist, said Garman died from oedema of the lungs following acute heart failure which took place during an attack of mania. He found no evidence of violence and he was satisfied that the drugs played no part in Garman's death. Mr Cal O'Callaghan, one of the male nurses, said when the struggle developed at the ambulance Garman shouted that he was being kidnapped. "The crowd was very threatening towards us and a woman caught hold of my sleeve. He was very strong and got back in the house," he said.

Dr. O'Connor was a friend of Kathleen's, one of a small group of younger Irish people whom she had befriended. Theo especially liked and trusted Desmond O'Connor, who owned six of his paintings. He was not a psychiatrist but an eye specialist who worked with cancer patients in a hospital just around the corner. (However, even psychiatrists did not know, then, that the drug probably put into Theo's food—sodium amytal—could cause the potentially fatal neuroleptic malignant syndrome, which may have been responsible for Theo's untimely death.) Theo often made notes in his books, sometimes in French. In one, he had written: "Des—je l'aime comme frère de sang" ("Des—I love him like a blood brother").

Perhaps the letter received by Chelsea police, referred to in this report, could have been the same letter that Helen alluded to once, in conversation with a friend. This friend said that she had asked Helen, who was so close to her sister, why on earth Kathleen went on sharing her home with Beth Lipkin, to which Helen had replied: "You see, it's all about the letter," before clamping her hand to her mouth and refusing to say more.

The press report certainly made sense of a troubling scene in a novel that its author, Wayland Young, had suggested I read. "It's called *Still Alive Tomorrow*," he had told me. "The reason I called it that was because so few of our friends were, at the time." As well as being close to Esther Garman before her suicide, Young and his wife were friends with the young cartoonist Mike Rutherston, who had so distressed Esther by killing himself after she refused his hasty proposal of marriage.

In *Still Alive Tomorrow*, Esther appears as Ruth, Mike Rutherston as Jimmy, and Esther's lover, Mark Joffe, is called Bela. Even Epstein, here called Rosenberg, makes an appearance, as does Ruth's powerful mother. At the center of the novel is the choice that the hero, Charles, makes between two women: dark, doomed Ruth and fair, pragmatic Claire. He chooses Claire but remains friends with Ruth, who duly comes to stay with him and his wife in their house in Italy. Here Ruth cries a lot and sits staring at the night sky and doesn't help with the washing up.

In the end, Ruth kills herself. But before this, there is a scene describing a terrible dinner party. Bela, like the real-life Joffe, is extremely possessive of his beautiful girlfriend, making it impossible for her to see her friends, even her family. In the novel, the dinner features Bela, but it seems the events it draws upon are those which took place on the night Theo died. A dinner party is arranged in order to drug Bela's food so that he will go quietly to a private mental hospital. There is a doctor whom Bela likes and trusts. He is invited to the dinner, and it is he who provides the drug for the food. An ambulance is already waiting outside. But the subterfuge doesn't quite work: there isn't enough dope in the food, and Bela is still restless. He is too agitated to get into the ambulance, so the doctor goes back into the house and administers an injection. Bela is said to have vomited and got the food stuck in his windpipe. The doctor performs an emergency tracheotomy, and Bela is admitted, as Theo was, not to a mental hospital, but to the local general hospital. Unlike Theo, he survives.

Several contemporaries had suggested that I should contact Wayland Young, who had known the family well during the 1950s. A handsome young writer and political activist, Young was to become involved with the Campaign for Nuclear Disarmament, as well as with feminist issues. His was a distinguished family: his mother had been married to Scott of the Antarctic, and his half-brother was the well-known wildlife conservator Sir Peter Scott. (Young's official title is Lord Kennet, but he and his wife both use his former name.)

The Youngs live in Bayswater in a remarkable house redolent of a tranquil, rather faded English country manor, despite the thundering traffic without. It was once the home of J. M. Barrie. I visited on a hot summer day, and we talked in the garden. The Youngs remembered Kathleen very clearly. They both thought she had been quite a difficult mother—"Certainly she was a very odd mother," said Elizabeth Young—expecting her children to conform to her own values, expecting them to have artistic talent. "She was a bit of a snob about artistic achievement," explained Elizabeth. She described Kathleen's voice as "slightly cawing; a bit rookish."

Elizabeth Young was warm and candid and obviously highly intelligent. She said that her husband had found it difficult to choose between herself and Esther, and that in order to resolve things, she had taken her rival for a drive around London. (As Kathleen once had, with Mrs. Epstein.) They had gone round and round Hyde Park Corner, deep in conversation; in the end they'd agreed that Elizabeth should have him. This resolved, they had all remained friends.

Esther had been at 272 on the night her brother died, and seen it all. Not long afterward, she went to stay with the Youngs in Italy, where they were then living. Esther had been very upset. It was when she came back to London from Italy that she first attempted to kill herself. The Youngs sensed that Esther, burdened by an overdeveloped sense of duty, had been a doomed figure. "She had a very strong sense of duty," said Elizabeth. "That's right," her husband acknowledged. "She was shackled by it. She was very unfree." All of us were silent for a time before Wayland Young spoke again: "In that family," he said, "there was a chair for suicide by the hearth, long before anyone occupied it."

Before I left, Wayland suggested I read his novel, and Elizabeth showed me the poem she had written about Esther, entitled "Suicide":

The darkness moved in too fast for her.
Wind, agitating her dark hair,
Blew up a night before her sun
Had reached its tolerable zenith.
Dark waves lapped at her.
Still radiant, she swam in them,
And what we loved her for was buoyancy
In the unhope she moved in;
Unhope she breathed too long.
A cavernous gale

The Rare and the Beautiful

Engulfed her; God withdrew
Perspective, humbled her insignificance.
She dropped back into her element.

Esther swimming in Italy in 1954.

Someone who had known Kathleen hinted that she may have wanted her son to die; that his life had become so difficult as to be untenable. He referred to a case that was then in the news, in which, in an act of desperate compassion, a father had helped his severely depressed daughter to end her life. Kathleen's acquaintance thought that drugging Theo's food had been a similar undertaking.

There is no evidence to suggest that Theo's death was some sort of mercy killing. My own view is that drugging his food was intended only to get him to go quietly back to the hospital. Theo's mental health was faltering, and his mother believed that a period in the hospital would help him. She wanted him to get better. With Epstein alone at Hyde Park Gate, needing her, she must have felt torn between her son and his father. If Kathleen was at fault, it was because she thought she could manage things; that she could do things, as she always had, without the rule book which other people live by. She could charm people, persuade them to do what she wanted. Young Dr. O'Connor would oblige her because people did. She had always lived like that. On that night, it simply went wrong.

But she did grieve, and she never stopped grieving. In a notebook she wrote:

Here is where I long to be
Where my darlings buried lie
Sheltered in security
Beneath the vaulted sky.
Sun and moon and stars above
Shine down eternally
On those I most on earth did love
Who now in earth do lie.
I kiss the earth above their head
And sing their lullaby.
The sweet flowers weave around the bed
Where I too hope to lie.

Every year for the rest of her life, she lit candles for her dead children on their birthdays.

Sources

Preface

Interviews

Stephen Gardiner, Kitty Garman, Dr. Sebastian Garman, Francis Wishart.

Books

McGregor, Sheila. 1999. *A Shared Vision*. London: Merrell Holberton.

Documents

Epstein, Kathleen. Unpublished letters. New Art Gallery, Walsall.
Godley, Kitty. Unpublished letter. Private collection (Cressida Connolly).

Chapter 1 The Black Country

Interviews

Deborah Chattaway, Yasmin David, Daphne Garman, Kitty Garman, Reggie Garman, William Garman, Barney Hutton, Dr. Philip Hutton, Marjorie Steward.

Books

Benson, John, and Trevor Raybould. 1978. *Walsall as It Was*. Nelson: Hendon Publishing.

Greenslade, M. W., and D. G. Stuart. 1965. *A History of Staffordshire.* Beaconsfield: Darwen Finlayson.

Grove, Valerie. 1999. *Laurie Lee.* London: Viking.

Documents

Epstein, Kathleen. Unpublished letters. New Art Gallery, Walsall.

————. Unpublished fragment of childhood memoir. Douglas Garman papers. Manuscripts and Special Collections, University of Nottingham Library.

Garman, Mavin. Unpublished memoir. In the possession of Dr. Sebastian Garman.

Garman, Walter; his mother, Mrs. Garman; and Marjorie Garman. Unpublished letters. In the possession of Kitty Garman.

Green, Arthur Romney. Unpublished memoir. National Art Library, Victoria and Albert Museum.

Powys, Llewelyn. Unpublished letter. Beinecke Rare Book and Manuscript Library, Yale University.

CHAPTER 2 LONDON

Interviews

Deborah Chattaway, Dr. Valerie Cowie, John Lade, Peggy Jean Lewis, Luke Wishart.

Books

Alexander, Peter. 1982. *Roy Campbell.* Oxford: Oxford University Press.

Campbell, Roy. 1934. *Broken Record.* London: Boriswood.

————. 1951. *Light on a Dark Horse.* London: Hollis & Carter.

David, Hugh. 1988. *The Fitzrovians.* Sevenoaks: Sceptre.

Deghy, Guy, and Keith Waterhouse. 1955. *Café Royal.* London: Hutchinson.

Gardiner, Stephen. 1992. *Epstein: Artist against the Establishment.* London: Michael Joseph.

Gray, Cecil. 1948. *Musical Chairs, or, Between Two Stools.* London: Home & Van Thal.

Hooker, Denise. 1986. *Nina Hamnett.* London: Constable.

John, Romilly. 1975. *The Seventh Child.* London: Jonathan Cape.

Lewis, Wyndham. 1937. *Blasting and Bombardiering.* London: Eyre & Spottiswoode.

Rose, June. 2002. *Daemons and Angels.* London: Constable.

Taylor, John Russell. 1990. *Bernard Meninsky.* Bristol: Redcliffe.

Documents

Lyle, Anna Campbell. Unpublished memoir. Private collection (Cressida Connolly).

Sources

CHAPTER 3 SUNSETS ETC

Interviews

Yasmin David, Anne Dunn, Rona Flack, Kitty Garman, Reggie Garman, Dr. Sebastian Garman, Dr. Philip Hutton, Francis Wishart.

Books

Alexander, Peter. 1982. *Roy Campbell.* Oxford: Oxford University Press.

Gardiner, Stephen. 1992. *Epstein: Artist Against the Establishment.* London: Michael Joseph.

Hale, Kathleen. 1994. *A Slender Reputation.* London: Frederick Warne.

Jones, J. D. F. 2001. *Storyteller.* London: John Murray.

Kiernan, Kathleen, Hilary Land, and Jane Lewis. 1998. *Lone Motherhood in Twentieth-Century Britain.* Oxford: Clarendon Press.

MacNeice, Louis. 1966. *Collected Poems.* London: Faber & Faber.

Pearce, Joseph. 2001. *Bloomsbury and Beyond.* London: HarperCollins.

Rose, June. 2002. *Daemons and Angels.* London: Constable.

Documents

Garman, Mavin. Unpublished memoir. Douglas Garman papers. University of Nottingham Library.

Garman, Kitty. Unpublished letters. Private collection (Cressida Connolly).

Plant, Wilson. Unpublished audiotape. Lady Beatrice Stowe and family.

Plomer, William. Papers. Durham University Library.

The Times, 19th October, 1923. The British Library, Colinwood Collection.

CHAPTER 4 MAGAZINES

Interviews

Kitty Garman.

Books

Alexander, Peter. 1982. *Roy Campbell.* Oxford: Oxford University Press.

Jones, J. D. F. *Storyteller.* London: John Murray.

Hobday, Charles. 1989. *Edgell Rickword.* Manchester: Carcanet.

Lucas, John. 1987. *The Radical Twenties.* Nottingham: Five Leaves.

Pearce, Joseph. 2001. *Bloomsbury and Beyond.* London: HarperCollins.

Documents

Campbell, Mary. Unpublished letters to William Plomer. Plomer Papers. Durham University Library.

Campbell, Roy. Unpublished account. Plomer Papers. Durham University Library.

Garman, Mavin. Unpublished memoir. Private collection (Dr. Sebastian Garman).

Bradbury, Malcolm. October 1961. *The Calendar of Modern Letters. London* magazine.

CHAPTER 5 THE SUMMER SCHOOL OF LOVE

Books

Campbell, Roy. 1949–60. *The Collected Poems of Roy Campbell.* London: Bodley Head.

Glendinning, Victoria. 1983. *Vita.* London: Weidenfeld & Nicolson.

Nicolson, Nigel. 1973. *Portrait of a Marriage.* London: Weidenfeld & Nicolson.

Sackville-West, Vita. 1929. *King's Daughter.* London: Hogarth Press.

Documents

Campbell, Mary. Unpublished letters to William Plomer. Plomer Papers. Durham University Library.

Nicolson, Harold. Unpublished letter. Plomer Papers. Durham University Library.

Sackville-West, Vita. "Liberty." *Harper's Bazaar,* October 1930.

CHAPTER 6 WING MY HEART

Interviews

Deborah Chattaway, Valerie Cowie, Rona Flack, Stephen Gardiner, Kitty Garman, Dr. Sebastian Garman, William Garman, John Lade, Niall Lawlor, Kathy Lee, Rosie Price.

Books

Alexander, Peter. 1982. *Roy Campbell.* Oxford: Oxford University Press.

Babson, Jane F. [c. 1984]. *The Epsteins.* Aylesbury: Taylor Hall.

Gardiner, Stephen. 1992. *Epstein.* London: Michael Joseph.

McGregor, Sheila. *A Shared Vision.* Introduction by Kitty Godley. London: Merrell Holberton.

Mitford, Nancy. 1945. *The Pursuit of Love.* London: Hamish Hamilton.

Nicholson, Virginia. 2002. *Among the Bohemians.* London: Viking.

Pearce, Joseph. 2001. *Bloomsbury and Beyond.* London: HarperCollins.

Rose, June. 2002. *Daemons and Angels.* London: Constable.

Wishart, Michael. 1977. *High Diver.* London: Blond & Briggs.

Documents

Epstein, Jacob. Letters to Kathleen Garman. New Art Gallery, Walsall.
Garman, Walter. Unpublished letters. Collection of Kitty Godley.
Godley, Kitty. Unpublished letters. Private collection (Cressida Connolly).

CHAPTER 7 MARTIGUES

Interviews

Deborah Chattaway, Kathy Lee, Anna Campbell Lyle.

Books

Alexander, Peter. 1982. *Roy Campbell*. Oxford: Oxford University Press.
Campbell, Roy. 1934. Broken Record. London: Boriswood.
———. 1951. *Light on a Dark Horse*. London: Haliz & Carter.
Hewitt, Kathleen. 1945. *The Only Paradise*. London: Jarrolds.
Hillier, Tristram. 1954. *Leda and the Goose*. London: Longmans, Green.
Holroyd, Michael. 1996. *Augustus John*. London: Vintage.
Lewis, Wyndham. 1932. *Snooty Baronet*. Ed. Bernard Lafourcade. Santa Barbara: Black Sparrow Press.
Pearce, Joseph. 2001. *Bloomsbury and Beyond*. London: HarperCollins.
Unterecker, John. 1970. *Voyager*. London: Anthony Blond.

Documents

Lyle, Anna Campbell. Unpublished memoir. Private collection (Cressida Connolly).
Campbell, Mary. Unpublished letters to William Plomer. Plomer Papers.
Durham University Library.

CHAPTER 8 PEGGY AND THE PARTY

Interviews

Anne Dunn, Kitty Garman, Dr. Sebastian Garman, Eric Hobsbawm, Sid Kaufman.

Books

Gill, Anton. 2002. *Peggy Guggenheim*. London: HarperCollins.
Guggenheim, Peggy. 1980. *Out of This Century*. London: André Deutsch.

Documents

Garman, Douglas. Papers. University of Nottingham Library.
Slater, Montagu. Papers. University of Nottingham Library.

Internet sources:
   Kevin Morgan and Andrew Flinn of the Communist Party of Great Britain
   Biographical Project at the University of Manchester. www.man.ac.uk

Chapter 9 You Beautiful Creature

Interviews
   Yasmin David, Rona Flack, Dr. Sebastian Garman, Kathy Lee, Pauline Tennant,
   Luke Wishart.

Books
   Graves, Richard Perceval. 1984. *The Brothers Powys*. Oxford: Oxford University Press.
   Grove, Valerie. 1999. *Laurie Lee*. London: Viking.
   Lee, Laurie. 1992. *As I Walked Out One Midsummer Morning*. London: Viking.
   Marlow, Louis. 1971. *Welsh Ambassadors*. London: Bertram Rota.

Documents
   Powys, Llewelyn. Unpublished letters. Beinecke Collection, Yale University
   Library.

Internet sources
   Judith Stinton of the Dorset County Museum. www.dorsetcountymuseum.org.
   Peter Foss of the Powys Society. www.powys-society.org.

Chapter 10 The Poet and the Painter

Interviews
   Johnny Craxton, Yasmin David, Kitty Garman, John Richardson, Pauline
   Tennant.

Books
   Feaver, William. 2002. *Lucian Freud*. London: Tate Gallery Publications.
   Powys, Llewelyn. 1950. *Love and Death*. London: Bodley Head.
   Russell, John. 1974. Introduction to catalogue, *Lucian Freud*. London: Arts
   Council of Great Britain.
   ———. 1993. *Lucian Freud*. New York: Robert Miller Gallery.

Articles
   Feaver, William. June 2002. Lucian Freud: the Unseen Pictures. *The World of
   Interiors*.

————. May 2002. Lucian Freud, Portraits of the Human Soul. *Guardian Weekend.*

Chapter 11 Lady Epstein

Interviews

Ruth Conybeare, Valerie Cowie, Jackie Epstein, Stephen Gardiner, Kitty Garman, William Garman, Roland Joffe, John Lade, Niall Lawlor, Kathy Lee, Rosie Price, June Rose, Susan Small, Hilary Ward, Kathleen Walne Ward, Elizabeth Young, Wayland Young.

Books

Gardiner, Stephen. 1992. *Epstein.* London: Michael Joseph.
Rose, June. 2002. *Daemons and Angels.* London: Constable.

Documents

Epstein, Jacob. Letter to Peggy Jean Lewis. Tate Gallery Archive.
Gayford, Martin. July 25, 1998. More Than a Face to Remember: Isabel Rawsthorne. *Daily Telegraph,* London.
Vivante, Cesare, and Professor Paolo Vivante. Unpublished letters to the author.

Chapter 12 Our Winter Season

Interviews

Kitty Garman, Anna Campbell Lyle.

Books

Alexander, Peter. 1982. *Roy Campbell.* Oxford: Oxford University Press.
Bedford, Sybille. 1973–4. *Aldous Huxley.* 2 vols. London: Chatto & Windus.
Campbell, Roy. 1949–60. *The Collected Poems of Roy Campbell,* 3 vols. London: Bodley Head.
Lee, Laurie. 1991. *A Moment of War.* London: Viking.
Pearce, Joseph. 2001. *Bloomsbury and Beyond.* London: HarperCollins.
Wishart, Michael. 1977. *High Diver.* London: Blond & Briggs.

Documents

Alexander, Peter. Unpublished letter to the author. Private collection (Cressida Connolly).
Lyle, Anna Campbell. Unpublished memoir.

CHAPTER 13 DEALING WITH DREAMS

Interviews

    Valerie Cowie, Roland Joffe, John Lade, Samuel Menashe, Rosie Price, Wayland
    Young.

Books

    Gardiner, Stephen. 1992. *Epstein*. London: Michael Joseph.
    McGregor, Sheila. 1999. *A Shared Vision*. London: Merrell Holberton.
    Rose, June. 2002. *Daemons and Angels*. London: Constable.

Documents

    Epstein, Kathleen. Unpublished letters to Beth Lipkin. Archive of the New Art
    Gallery, Walsall.
    Material about the Little Gallery. Archive of the New Art Gallery, Walsall.
    A copy of Sally Ryan's will and outline of her family and biography are in the
    archive of the New Art Gallery, Walsall.

CHAPTER 14 NOTHING IN MODERATION

Interviews

    Vicky Bryson, Deborah Chattaway, Julian David, Yasmin David, Anne Dunn,
    Rona Flack, Kitty Garman, Dr. Sebastian Garman, William Garman, Fernanda
    Gazdar, Eric Hobsbawm, Sid Kaufman, Pauline Tennant, Francis Wishart, Luke
    Wishart, Michelle Wishart.

Books

    Breese, Charlotte. 1999. *Hutch*. London: Bloomsbury.
    Grove, Valerie. 1999. *Laurie Lee*. London: Viking.
    Hobsbawm, Eric. 2002. *Interesting Times*. London: Allen Lane.

Documents

    Garman, Douglas. Papers. Manuscripts and Special Collections. Nottingham
    University Library.
    Wilson, Jeremy. Unpublished e-mail. Private collection (Cressida Connolly).
    Wishart, Lorna. Tape recording. Private collection (Vicky Bryson).

APPENDIX THEO AND ESTHER

Interviews

Dr. Chris Beyrer, Samuel Menashe, Gerard Ogle, Dr. Peter Sanderson, Sydney Smith, Elizabeth Young, Wayland Young.

Books

McGregor, Sheila. 1999. *A Shared Vision*. London: Merrell Holberton.
Young, Wayland. 1958. *Still Alive Tomorrow*. London: Cresset Press.

Documents

London Evening News. 1954. "Split Mind" Artist in Ambulance Struggle. February 1.

BIBLIOGRAPHY

Ackroyd, Peter. *T. S. Eliot.* London: Penguin Books, 1993.

Alexander, Peter. *Roy Campbell: A Critical Biography.* Oxford: Oxford University Press, 1982.

Ariès, Phillippe. *Centuries of Childhood.* Translated by Robert Baldick. London: Pimlico, 1996.

Babson, Jane F. *The Epsteins: A Family Album.* Aylesbury: Taylor Hall, [c. 1984].

Bedford, Sybille. *Aldous Huxley: A Biography.* 2 vols. London: Chatto & Windus, 1973–4.

Beekman, Daniel. *The Mechanical Baby: A Popular History of the Theory and Practice of Child Raising.* London: Dobson, 1979.

Benson, John, and Trevor Raybould. *Walsall as It Was.* Nelson: Hendon Publishing, 1978.

Borodin, George. *This Thing Called Ballet.* London: Macdonald & Co., 1946.

Bradbury, Malcolm, "*The Calendar of Modern Letters,*" *London Magazine,* October 1961.

Brain, W. Russell. *Some Reflections on Genius and Other Essays.* London: Pitman Medical Publishing, 1960.

Breese, Charlotte. *Hutch.* London: Bloomsbury, 1999.

Buckle, Richard. *Jacob Epstein: Sculptor.* London: Faber & Faber, 1963.

Buckman, David. *Mixed Palette: The Painting Lives of Frank Ward and Kathleen Walne.* Bristol: Sansom & Co., 1997.

Campbell, Anna. *Poetic Justice.* Francestown: Typographeum, 1986.

Campbell, Roy. *Broken Record: Reminiscences.* London: Boriswood, 1934.

————. *The Collected Poems of Roy Campbell.* 3 vols. London: Bodley Head, 1949–60.

————. *Flowering Rifle: A Poem from the Battlefield of Spain.* London: Longmans & Co., 1939.

————. *Light on a Dark Horse: An Autobiography 1901–1935.* London: Hollis & Carter, 1951.

————. *Sons of the Mistral.* London: Faber & Faber, 1941.

Carruthers, Annette, and Mary Greensted. *Good Citizen's Furniture: The Arts and Crafts Collection at Cheltenham.* Cheltenham: Cheltenham Art Gallery and Museum, 1994.

Crick, Bernard. *George Orwell: A Life.* London: Secker & Warburg, 1980.

Croft, Andy, ed. *A Weapon in the Struggle: The Cultural History of the Communist Party in Britain.* London: Pluto Press, 1998.

Daintrey, Adrian. *I Must Say.* London: Chatto & Windus, 1963.

David, Hugh. *The Fitzrovians: A Portrait of Bohemian Society 1900–1955.* Sevenoaks: Sceptre, 1988.

Davies, V. L., and H. Hyde. *Dudley and the Black Country 1760–1860.* Dudley Public Libraries Transcript No. 16. Dudley: County Borough of Dudley, Libraries, Museums and Arts Department, 1970.

Deghy, Guy, and Keith Waterhouse. *Café Royal: Ninety Years of Bohemia.* London: Hutchinson, 1955.

Ede, John F. *History of Wednesbury.* Wednesbury: Wednesbury Corporation, 1962.

Elkin, Susan. *Life to the Less: A Biography of Arthur Romney Green.* Christchurch: Natula, 1998.

Elwin, Malcolm. *The Life of Llewelyn Powys.* London: John Lane at the Bodley Head, 1946.

Epstein, Jacob. *Let There Be Sculpture: An Autobiography.* London: Michael Joseph, 1940.

Feaver, William. *Lucian Freud.* London: Tate Gallery Publications, 2002.

Fletcher, Ronald. *The Family and Marriage in Britain: An Analysis and Moral Assessment.* Harmondsworth: Penguin Books, 1966.

Gallagher, Jock, ed. *Laurie Lee: A Many-Coated Man.* London: Viking, 1998.

Gardiner, Stephen. *Epstein: Artist against the Establishment.* London: Michael Joseph, 1992.

Gertler, Mark. *Selected Letters.* Edited by Noel Carrington. London: Rupert Hart-Davis, 1965.

Gill, Anton. *Peggy Guggenheim: The Life of an Art Addict.* London: HarperCollins, 2002.

Glendinning, Victoria. *Vita: The Life of V. Sackville-West.* London: Weidenfeld & Nicolson, 1983.

Graves, Richard Perceval. *The Brothers Powys.* Oxford: Oxford University Press, 1984.

Gray, Cecil. *Musical Chairs, or, Between Two Stools, Being the Life and Memoirs of Cecil Gray.* London: Home & Van Thal, 1948.

Greenslade, M. W., and D. G. Stuart. *A History of Staffordshire.* Beaconsfield: Darwen Finlayson, 1965.

Gregory, Alyse. *The Cry of a Gull: Journals 1923–1948*. Edited by Michael Adam. Dulverton: Ark Press, 1973.

Grove, Valerie. *Laurie Lee: The Well-Loved Stranger*. London: Viking, 1999.

Guggenheim, Peggy. *Out of This Century: Confessions of an Art Addict*. London: André Deutsch, 1980.

Hamnett, Nina. *Laughing Torso*. London: Constable, 1932.

Hale, Kathleen, *A Slender Reputation: An Autobiography*. London: Frederick Warne, 1994.

Hewitt, Kathleen. *The Only Paradise: An Autobiography*. London: Jarrolds, 1945.

Hillier, Tristram. *Leda and the Goose: An Autobiography*. London: Longmans, Green & Co., 1954.

Hoare, Phillip. *Noël Coward: A Biography*. London: Sinclair-Stevenson, 1995.

Hobday, Charles. *Edgell Rickword: A Poet at War*. Manchester: Carcanet, 1989.

Hobsbawm, Eric. *Interesting Times: A Twentieth-Century Life*. London: Allen Lane, 2002.

Holroyd, Michael. *Augustus John: The New Biography*. London: Vintage, 1996.

Hooker, Denise. *Nina Hamnett, Queen of Bohemia*. London: Constable, 1986.

Jackson, Stanley. *An Indiscreet Guide to Soho*. London: Muse Arts, 1946.

Jenkins, Alan. *The Thirties*. London: Heinemann, 1976.

———. *The Twenties*. London: Heinemann, 1974.

John, Romilly. *The Seventh Child: A Retrospect*. London: Jonathan Cape, 1975.

Jones, J. D. F. *Storyteller: The Many Lives of Laurens van der Post*. London: John Murray, 2001.

Kiernan, Kathleen, Hilary Land, and Jane Lewis. *Lone Motherhood in Twentieth-Century Britain: From Footnote to Front Page*. Oxford: Clarendon Press, 1998.

Lee, Laurie. *Red Sky at Sunrise. Cider With Rosie. As I Walked Out One Midsummer Morning. A Moment of War*. London: Viking, 1992.

Lewis, Wyndham, *Snooty Baronet*. Edited by Bernard Lafourcade. Santa Barbara: Black Sparrow Press, 1984.

Lucas, John. *The Radical Twenties: Aspects of Writing, Politics and Culture*. Nottingham: Five Leaves, 1997.

Marlow, Louis. *Welsh Ambassadors: Powys Lives and Letters*. London: Bertram Rota, 1971.

McGregor, Sheila. *A Shared Vision: The Garman Ryan Collection at the New Art Gallery, Walsall*. With an introduction by Kitty Godley and commentaries by Oliver Buckley. London: Merrell Holberton, 1999.

Menashe, Samuel. *Fringe of Fire: Poems*. London: Victor Gollancz, 1973.

Mitford, Nancy. *The Pursuit of love*. London: Hamish Hamilton, 1945.

Nelson, Michael. *A Room in Chelsea Square*. London: Jonathan Cape, 1958.

Nicholson, Virginia. *Among the Bohemians: Experiments in Living*. London: Viking, 2002.

Bibliography

Nicolson, Nigel. *Portrait of a Marriage*. London: Weidenfeld & Nicolson, 1973.

O'Keeffe, Paul. *Some Sort of Genius: A Life of Wyndham Lewis*. London: Jonathan Cape, 2000.

Pearce, Joseph. *Bloomsbury and Beyond: The Friends and Enemies of Roy Campbell*. London: HarperCollins, 2001.

Plomer, William. *Celebrations*. London: Jonathan Cape, 1972.

Powys, Llewelyn. *The Letters of Llewelyn Powys*. Edited by Louis Wilkinson. Introduction by Alyse Gregory. London: John Lane, 1943.

———. *Love and Death: An Imaginary Autobiography*. London: Bodley Head, 1950.

Richardson, John. *Sacred Monsters, Sacred Masters: Beaton, Capote, Dalí, Picasso, Freud, Warhol and More*. London: Jonathan Cape, 2001.

Rose, June. *Daemons and Angels: A Life of Jacob Epstein*. London: Constable, 2002.

Rothenstein, John. *Summer's Lease: Autobiography, 1901–1938*. London: Hamish Hamilton, 1965.

Russell, John. Introduction to catalogue. *Lucian Freud*. London: Arts Council of Great Britain, 1974.

———. *Lucian Freud: Early Works*. New York: Robert Miller Gallery, 1993.

Sackville-West, Vita. *King's Daughter*. London: Hogarth Press, 1929.

Shulman, Nicola. *A Rage for Rock Gardening: The Story of Reginald Farrer, Gardener, Writer and Plant Collector*. London: Short Books, 2002.

Silber, Evelyn. *The Sculpture of Epstein*. Oxford: Phaidon, 1986.

Sinclair, Andrew. *War Like a Wasp: The Lost Decade of the Forties*. London: Hamish Hamilton, 1989.

Sitwell, Edith. *Selected Letters of Edith Sitwell*. Edited by Richard Greene. London: Virago, 1997.

———. *Taken Care Of: An Autobiography*. London: Hutchinson, 1965.

Stevenson, John. *British Society 1914–45*. London: Penguin Books, 1990.

Strinati, Dominic. *An Introduction to Theories of Popular Culture*. London: Routledge, 1995.

Sylvester, David. *About Modern Art: Critical Essays 1948–2000*. London: Pimlico, 2002.

Taylor, John Russell. *Bernard Meninsky*. Bristol: Redcliffe, 1990.

Tristram, Emma, ed. *Binsted and Beyond*. Sussex: Friends of Binsted Church, 2002.

Unterecker, John. *Voyager: A Life of Hart Crane*. London: Anthony Blond, 1970.

West, Alick. *One Man in His Time: A Personal Story of This Revolutionary Century*. London: Allen & Unwin, 1969.

Wishart, Michael. *High Diver*. London: Blond & Briggs, 1977.

Young, Wayland. *Still Alive Tomorrow*. London: Cresset Press, 1958.

Acknowledgments

It was visits to my friend, the Ladybird artist Harry Wingfield, which first brought me to Walsall and the Garmans. I remember Mr. Wingfield with gratitude, affection, and respect.

Thank you to my husband, Charles, and to Violet, Nell, and Gabriel for allowing me the time to think about, then write, this book.

And thanks, also, to Deirdre Levi, Deborah Orr, and, especially, to Catherine Palmer for their thoughts about the appendix.

PHOTOGRAPH CREDITS

The Garman siblings in 1913, page 9. Courtesy of the New Art Gallery, Walsall. Kathleen Garman, circa 1920, page 16. Courtesy of the New Art Gallery, Walsall. Kathleen Garman around the time she met Jacob Epstein, 1921, page 25. Courtesy of the New Art Gallery, Walsall. Sylvia and Lorna Garman in Herefordshire, 1924, page 45. Courtesy of Mrs. Julian David. Douglas Garman in his final year at Cambridge, page 54. Courtesy of The Beth Lipkin Archive at the New Art Gallery, Walsall. Douglas and Jeanne Garman with baby Deborah, 1926, page 70. Courtesy of The Beth Lipkin Archive at the New Art Gallery, Walsall. Mary Campbell with baby Anna and toddler Tess in Kent, 1927; page 81. Courtesy of The Beth Lipkin Archive at the New Art Gallery, Walsall. Rosalind, Kathleen with baby Theo, and Helen, spring 1925, page 103. Courtesy of The Beth Lipkin Archive at the New Art Gallery, Walsall. Lorna Wishart with baby Michael, circa 1928, page 124. Courtesy of Lady Beatrice Stowe. Mary Campbell in the south of France, early 1930s, page 126. Courtesy of the New Art Gallery, Walsall. Jeanne Hewitt Garman, circa 1932, page 136. Courtesy of Mrs. William Chattaway. Douglas Garman with Peggy Guggenheim and Helen Garman, 1936, page 157. Courtesy of Mrs. Laurie Lee. Studio portrait of Lorna Wishart, circa 1930, page 168. Courtesy of William Garman. Laurie Lee, Mary and Roy Campbell in Toledo, 1935, page 177. Courtesy of the late Anna Campbell Lyle. Lorna Wishart with Yasmin, winter 1941, page 186. Courtesy of Mrs. Julian David. Lucian Freud with zebra head, page 197. Reprinted from *Lucian Freud*, Tate Publishing, London, 2002. Kathleen Garman at the beach, circa 1945, page 213. Courtesy of The Beth Lipkin Archive at the New Art Gallery, Walsall. Theo Garman and Kathleen Garman in the garden of 272 King's Road, early 1950s, page 216.

INDEX